THE WEATHER REVOLUTION

Innovations and Imminent
Breakthroughs in
Accurate Forecasting

ALSO BY JACK FISHMAN AND ROBERT KALISH

GLOBAL ALERT
The Ozone Pollution Crisis

THE WEATHER REVOLUTION

Innovations and Imminent Breakthroughs in Accurate Forecasting

Jack Fishman
and
Robert Kalish

PLENUM PRESS • NEW YORK AND LONDON

Library of Congress Cataloging-in-Publication Data

Fishman, Jack, 1948-
    The weather revolution : innovations and imminent breakthroughs in
accurate forecasting / Jack Fishman and Robert Kalish.
        p.   cm.
    Includes bibliographical references and index.
    ISBN 0-306-44764-9
    1. Weather forecasting.   I. Kalish, Robert.   II. Title.
QC995.F55   1994
551.6'3--dc20   94-25380
                                                                     CIP

ISBN 0-306-44764-9

© 1994 Jack Fishman and Robert Kalish
Plenum Press is a Division of Plenum Publishing Corporation
233 Spring Street, New York, N.Y. 10013-1578

Printed in the United States of America

For
Mickey and Irv
and
Goldie and Simon

Preface

WHAT REVOLUTION ARE WE TALKING ABOUT?

Even the most down-to-earth person must wonder if Mother Nature has gone just a little bit out of control in the 1990s. Within a very brief period, the world has witnessed weather events of such unusual proportions that it's hard to believe that *something* has not happened to force all these occurrences upon our planet in such a short time span.

The drought that gripped southern and eastern Africa in 1992 was the worst in that region in more than a century. The storm that dumped snow from Florida to Maine and produced winds in excess of 100 miles per hour in March 1993 sent weather buffs scurrying to their archives to see when was the last time that such a powerful storm had ever hit the United States. They concluded that no other storm in this century possessed the strength that this one had. Then came the floods in July and August, 1993. Water levels reached heights that had never before been witnessed. Although many areas along the Mississippi were prepared to withstand flooding of magnitudes that should occur no more than once every hundred years, the upper Midwest faced a situation that some experts said should happen with a frequency of no more than once every *five hundred* years.

Nor were the heavily populated regions of the northeast United States immune from weather stories. By the 1993–1994 winter's end, tens of millions of Americans from Boston to Washington were crying, "Enough is enough!" as they suffered through storm after storm that broke long-standing snowfall records at many locations within this region.

When we started work on this book, Nature's onslaught was not the revolution we had in mind. We wanted to tell about all the technological advances that were reported to be "just around the corner." These included several new satellites, a new type of radar that could "see" wind structure within a thunderstorm, and the promise of even more exciting tools once the use of lasers placed in space started to send data back to earth with detail unimaginable just a few years ago.

The development of the new tools of the weather forecaster is related to an even more general revolution that is being experienced by the society we now live in—a society that has become extremely dependent on the use of computers to function normally. Some refer to this revolution as the "Information Age"—it has manifested itself in the business of weather presentation as well.

Just think about a weather broadcast today versus those before personal computers became so commonplace. The increased power of computers, along with their affordability, was the driving force that allowed the presentation of weather to become entertainment on a day-to-day basis. The TV weatherperson's colored markers and paste-up umbrellas were replaced by the computer mouse that drove the software that drew all these features much quicker and better than any person could. Lines on the weather chart were replaced by actual satellite pictures that had been "computer enhanced" so that you didn't need a Ph.D. to understand what was being seen from space.

Without this technological revolution, imagine how boring the Weather Channel would be. Then again, the Weather Channel most likely wouldn't even be on our cable networks if it weren't for the technological advancements that made weather entertainment possible.

There appear to be several revolutions currently taking place that are relevant to the times for which this book has been written. Revolutions generally do not happen overnight. In retrospect, we may view a particular event as a turning point, the moment that reshapes the lives of many people for decades to come. This book will guide the reader through some of these recent moments. We will also make you keenly aware of some of the exciting developments that will lead to events in the near future that in turn, may be seen as harbingers for the way all weather forecasts will be for many decades to come. Who knows? Perhaps someday we will look at an inaccurate forecast as a truly unusual happening.

Jack Fishman
Robert Kalish

Poquoson, Virginia
Bath, Maine

Acknowledgments

Many people helped in the preparation of this book. The list includes: Tom Schlatter, Ron Gird, G. V. Rao, Don Cahoon, Lou Uccellini, Vince Brackett, Toffee Fakhruzzaman, Mary Osborne, Henry Fuelberg, Mike Lewis, Lucien Froidevaux, Marta Fenn, Monica Gaines, Richard Eckman, Greg Stumpf, Doreen Nadolny, John Theon, and Paul Cousins. The thoughtful comments of Uccellini and Donald Johnson helped make this a much better book. We thank Linda Regan and Melicca McCormick, our editorial staff at Plenum, for their assistance. Last, but certainly not least, we thank our families, David, Jason, Melissa, Sam, and Will, and especially our mates, Sue and Eloise, for giving us their love and understanding during the many times when writing this book kept us from being with them.

Contents

THE WEATHER REVOLUTION

Innovations and Imminent
Breakthroughs in
Accurate Forecasting

ONE

Why Weather Forecasters Are So Often Wrong

It's been dubbed The Great Blizzard of '78—1978, that is. The Great Blizzard of February 6–7, 1978 totally shut down an area from New Jersey to Maine, dumping 20 inches of snow on the Boston area. It crippled the lives of more than 10 million people, who all had stories to tell. Some were forced to abandon their cars on the Central Artery, the expressway that cuts a swath through the center of the city, and had to take shelter at local schools; some whittled away the night at Logan Airport, playing rummy with other stranded passengers while intermittently sneaking peeks at the whitecaps out on the harbor. And there were those who thought they were safe and secure in their luxurious well-to-do homes on the South Shore, only to witness the awesome force of the sea as it reclaimed the beaches out from under them.

The 20-inch snowfall set a record for Boston. But that wasn't the worst of it. What created the most havoc was the wind. In the midst of the snow came a roaring wind out of the southeast that pushed before it a tidal surge of seawater.

The waves that pummeled the coast from Cape Cod to Maine left more damage in their wake than most hurricanes.

The snowstorm that hit the first week of February was the second major storm in less than two weeks. Unlike the storm that hit the Northeast on January 26, which dumped over a foot of snow, the February storm was well predicted. It came from the Midwest, hitching a ride on the strong flow of the jet stream and gaining energy along the way. But the January storm had surprised weather forecasters in New York and Boston. That storm had formed off the coast of North Carolina and forecasters had predicted it would turn out to sea before reaching the Northeast. But it didn't. Instead, it turned inland as it approached the Northeast and buried the region in blizzardlike conditions.

What was the difference between the two storms? Why was one prediction right on the mark and the other off? Despite the advance warning of the second storm's arrival (the National Weather Service had notified New York City officials a full day before the storm actually began), public officials were hesitant to take precautions because of the missed forecast of the previous storm. Yet the National Weather Service had done its job, it had alerted the public to a major storm.

The two snowstorms of 1978 are good examples of the complexity of modern weather forecasting. At the more mundane level, it would surprise no one to say that today, as we head into the twenty-first century, the National Weather Service's three-day and five-day forecasts are not as accurate as we would like. As a matter of fact, a three-day forecast in many areas of the country is only about 50% accurate. An average of .500 may be superlative for a professional baseball player, but not for weather forecasting: it means that the forecaster has "blown the forecast" three days into the future just about as many times as he's gotten it right!

And it's not just the eastern United States where forecasting is imperfect. An analysis of weather forecasting for ten summers ending in 1986 in Honolulu showed that forecasts were 83% correct. That sounds pretty good, until you realize that the climate in Honolulu is not like the climate anywhere else on the mainland. The climate in Honolulu is fairly uniform, a veritable tropical paradise. In fact, the climate in Honolulu is so uniform that a one-day forecast of no rain every day of the year would have been 85% correct. In other words, the forecasters in Honolulu did worse with their forecasts than no forecast at all.[1] So why is it so difficult to forecast accurately?

Weather forecasting for the public has caused trouble and confusion from the birth of modern meteorology. In 1854, the United Kingdom government created the Meteorological Department on the recommendation of the Royal Society, the British equivalent of the United States Academy of Sciences. The first head of the Department was Captain Robert Fitzroy, who was expected to collect data from ships and, because of the recent invention of the telegraph, create weather charts. From these charts, Fitzroy issued some of the first forecasts. In 1861, Fitzroy issued his first storm warnings for coastal regions, and shortly thereafter, the London *Times* began publishing his weather forecasts.[2]

Of course mistakes were made, and the scientists of the Royal Society criticized him for doing what he was supposed to do not very well. In his 1863 report, Fitzroy noted "certain persons [in the Royal Society] who were opposed to the system [of forecasting] theoretically at its origins," and "those who undervalue the subject [of weather forecasting] and to call it burlesque."[1] Although he tried to show that his forecasts were based on science, many members of the Royal Society and in the lay community continued to criticize the Meteorological Department's head.

Two years later Fitzroy, in apparent mental agony, committed suicide. Soon after, a special committee was formed to investigate the Meteorological Department and concluded: "The first step toward placing the prediction of storms on a clear and accurate basis was taken by Fitzroy. He compelled scientists and society in general to take note of this branch of meteorology, and he gave his life in the effort."[2] But this same committee also decided that "there is as yet no real scientific justification for daily predictions. The accuracy of such predictions has not in general been proven, and no data exist which demonstrate their usefulness."[3]

Soon after, the Meteorological Office was closed, to be reopened in December 1867 in response to public demand, and forecasts were resumed. More than a 100 years later we're still hearing many of the same criticisms directed at forecasters. Everyone has had the experience of hearing a television weather forecaster announce on Wednesday that "the upcoming weekend looks wonderful." Taking the weather forecaster at his word, families plan that picnic, that day at the beach, that outdoor birthday party with 30 children, only to have Saturday morning arrive sodden and gray, anything but glorious. Whom do we blame? The weather forecaster. At least now, the stress of an incorrect forecast isn't placed so clearly on the shoulders of one individual. And we don't see modern-day forecasters going the same route as Fitzroy because of a busted forecast.

Are forecasters to be blamed? Obviously, they don't

make the weather, they just try to predict it. Not that many years ago, when one of us was in school studying meteorology—more than 100 years after Fitzroy—many scientists still considered weather forecasting only half science, the other half art. To be truly accurate, wizened old weather forecasters knew that you had to incorporate an "artful" component into your scientific forecast. In other words, all the data, numbers, and mathematical formulas weren't enough for an accurate prediction. You needed an extra ingredient, and that's where the other half, the "art," entered the picture.

But why? Isn't weather forecasting strictly scientific? Don't forecasters use the latest in computers, satellites, laser beams, and other high-tech aids undreamed of just a short time ago? Yes, they do. And still, they can't predict weather more than three days into the future with complete confidence. Despite all the scientific advances that have occurred over the last 50 years, weather refuses to be controlled or submit to accurate prediction. It's a wild and unruly child, to be sure. Not particularly mean, only playful. A child listening to a different drummer.

And that, perhaps, is what makes the study of weather so fascinating. After all, isn't the weather, with its constantly changing, dynamic uncertainty, a reminder to all of us that we, too, are constantly changing and dynamic? Science has been pulling the rug out from beneath our conceptions of ourselves and our world. There was a time when the laws of physics supported the notion that matter was unchanging, that solid things were solid, gases were gases, and liquids were liquids. Some of the new theories, such as quantum physics, have changed all that. According to quantum physics, everything in the universe is in a constant state of flux. Nothing is as solid as it appears. Nothing is permanent.

Weather, then, with its ever-changing nature, with its indomitable ability to surprise, is a constant reminder that this world of ours is also in a state of constant flux despite our attempts to control it. There are aspects of our lives that we do manage to control. If we're fortunate, we have jobs, families, a place to live. These things help foster a sense of security, allowing us to go to sleep at night pretty sure that the next morning our jobs and homes and families will be waiting for us.

Weather is a reminder that we and our world are not as solid and substantial as we'd like to believe, that our idea of ourselves as permanent and unchanging is false. Weather, after all, is a manifestation of the movement of energy. Such movement, the flux and flow of energy, occurs all the time, every moment. Weather is a reminder that the earth is a living organism, always in the midst of a dynamic change.

At first glance, the scientific part of forecasting the weather doesn't seem to be that difficult. After all, atmospheric motions must follow certain laws of physics that meteorologists have known for years. Yet they still cannot describe these motions accurately enough to give us a truly reliable 24-hour forecast all the time, much less a five-day forecast that can be believed even half the time. One reason is that, unlike most other scientists, meteorologists cannot verify their calculations by making measurements in a controlled laboratory environment. An engineer, for example, can study how steel reacts under strong winds by observing a model steel airplane in a small wind tunnel. He can control the wind, the pressure, all the variables in his small laboratory. The results can then be translated to actual airplanes in actual wind. But meteorologists don't have an isolated laboratory to experiment in. Their laboratory is the atmosphere itself, the one right outside their window. They must look out the window, must study the atmosphere itself, to observe how it behaves, and thus to verify their calculations. In other words, they must observe the weather both to forecast it and to evaluate their previous forecasts.

What are these laws of physics that rule the atmosphere? We have to begin with the sun, the source of all the earth's energy. The sun's energy is intercepted by the earth, and so it warms the land and oceans. Because heat expands, the heating from the sun's energy at the bottom levels of the earth's atmosphere causes the fluid above it to move. You can see such a physical law at work when you put on a pot of water to boil. As the bottom level of water heats up, you can see the hot water begin to move, to lift. This is called convection, the tendency of fluids or liquids to rise as they are heated.

But the earth is a bit more complicated than a pot of water. For one thing, the earth's shape is such that the area near the equator receives more of the sun's energy than the areas toward the poles. Another factor is the diversity of the earth, the juxtaposition of land and water. Land surfaces heat more rapidly than water surfaces, causing a natural convection to occur. Once the convection is set in motion, the rotation of the earth deflects the movement of such currents. In addition there are a variety of topographics—mountains that rise tens of thousands of feet, deserts that lie far below sea level, canopied forests, cold lakes—that affect the way in which the atmospheric laws play out.

Still, nothing is new here. Scientists have known for decades about the effects of mountains on weather, and that oceans take longer to warm up and cool off than land surfaces. These parameters have remained constant for millennia. Such constants are the raw materials of a weather forecast, the known data. So, knowing these parameters, the question remains: why can't meteorologists make accurate forecasts? The answer is a complex one.

THE COMPUTER AGE

It's probably difficult for a modern meteorology student to understand how exciting it was twenty years ago when eager students stampeded into stores to buy the first hand-held calculators. The Hewlett-Packard HP-35 was a dream come true: no more slide rules and trig tables. Students back then had faith that technology would empower them to make accurate, long-range weather forecasts. At the time, they thought the primary hurdle was computer power. They had the raw data, they knew how weather was made, all they needed was the technology to carry out the calculations.

Twenty years later the world enjoys computer power unimagined back then. Yet the goal of accurate, long-range weather forecasts still eludes us. Why? Because in the ensuing twenty years, while computers have multiplied and grown in sophistication, our knowledge of weather dynamics has also grown and deepened. And we now also know that there's a lot more they need to know to produce perfect prognostications.

For example, twenty years ago scientists didn't realize just how important the oceans were to the formation of weather. They now know that much of the interannual variability—the fact that not every winter is as cold or as warm as the previous winter—is a consequence of how the heat captured in the oceans interacts with the atmosphere from one year to the next. Computers and satellites have helped us gain this kind of knowledge, but it seems the more we learn the more there is to learn.

Knowledge of the oceans and their effect on the atmosphere has helped meteorologists understand long-term weather behavior that at first appears unconnected. For example, we've learned that the phenomenon known as El Niño, a warming of the ocean off the coast of Peru, profoundly affects the extent and duration of droughts and the number and intensity of hurricanes on North America's east coast.

They've done better with severe kinds of weather too. The development of Doppler radar, for example, allows them to make much more accurate estimates of where tornadoes will break out. We can track hurricanes for weeks and prepare land-based populations for the worst. Some advances have been made in technology, and in the application of these new tools. But the bottom line is still the weather forecast for the coming weekend, for the next day. And despite the new technology and understanding, we are too often confronted with a "busted" forecast for such routine predictions.

But this too is changing. Within this decade, by the turn of the century, scientists will have made a quantum leap in the ability to "see" the atmosphere, and thus to make accurate, long-range forecasts. New technology will make it possible to measure the atmosphere far more accurately than is possible now.

The way weather forecasts are compiled today may be likened to the old days of the automobile. We are still in the early days of the Model A and Model T. To be sure, we have modern computers, but what are we putting into them? Computers are only as good as the data entered into them. And this is where the change will occur first.

Most of the data fed into the computer for a weather forecast are obtained by releasing slow-rising balloons to measure atmospheric parameters. In other words, despite the sophisticated numerical models the computer can give us, a large percentage of the data we give the computer are obtained by the same means used nearly a century ago.

But there are several new tools on the horizon that will probe the atmosphere in ways that were unthinkable in the 1950s when computers hit the scene and revolutionized the science of weather forecasting as we generally know it today. One of these tools is the laser. Laser beams are able to make measurements at the speed of light, but the computer technology needed to process these measurements is just now emerging.

Balloons are released at certain land-based stations regularly. As they rise through the atmosphere, instruments on board record such data as temperature, humidity, and pressure. We can track these balloons as they rise and determine how the wind is blowing. The problem is that such information, while complete, is only accurate for the thin vertical window traveled by the balloon. With the laser technology just over the horizon, we will be able to obtain the same readings at the speed of light, and gain such information from a much broader area, including the atmosphere over the oceans, as well as from the most remote to the most densely populated areas.

It is such new laser-equipped satellites that will be at the heart of the coming revolution in weather forecasting. Instead of waiting for a balloon to drift upward for ten miles, laser beams from far above the stratosphere will probe the atmosphere and feed us information instantaneously.

Figure 1 helps illustrate what the use of such satellites will mean to your local forecaster. This figure shows the existing upper-air network over North America. The upper-air network consists of special weather stations that rou-

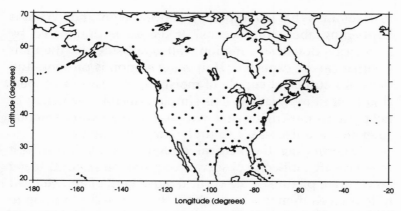

FIGURE 1. Location of upper-air stations. The primary way of gathering weather data for use in forecasts is by means of balloons. This map shows the locations from which rawinsondes are launched twice daily. As these balloons rise through the atmosphere, they radio back information about the temperature, humidity, and pressure structure to the atmosphere. In addition, the vertical distribution of winds can be calculated by tracking them by radar as they rise. Note that very little information is obtained over the oceans (70% of the earth's surface is covered by the oceans).

tinely launch weather balloons to measure temperature, humidity, pressure, and winds throughout its vertical path through the atmosphere.

There are far fewer of these upper-air stations than there are surface stations, which record temperature, humidity, pressure, and winds at the surface. Yet these upper-air stations are important because forecasters need the information about the upper atmosphere to generate accurate forecasts for the surface.

When we look at a weather chart, we see only two dimensions. But the real key to understanding and forecasting the weather is that third dimension, the vertical movement of air. What is happening above the ground? What is happening above the oceans? If the air is dominated by large-scale downward motion, called subsidence, then fair weather can be expected. If upward motion is present, and there is a sufficient supply of moisture, then we can expect cloudy or rainy weather. This simple principle—whether or not the air is moving up or down—is one of the fundamental rules that are necessary to understand the weather.

Determining the vertical direction of air is difficult when we are limited to observations from the ground. To get an accurate picture of the vertical movement of air, we need to take measurements throughout its vertical domain, up to 40,000 or 50,000 feet. That's generally the height limit of the upper level of the troposphere, where weather takes place. Up until now, we've taken such measurements with balloons only at the upper-air stations. At such stations in the United States, the balloons are launched twice a day.

There are only a few regions in the world where upper-air networks are as numerous as they are in the United States. Outside the United States, Europe, and Japan, such upper-air networks are scarce, and they launch balloons only once a day when they operate at all. If you consider the vastness of the earth's atmosphere, such widely scattered upper-air stations can't help but fail at gathering enough data. One reason there aren't more stations is money.

The balloons that are sent up to record data are called rawinsondes (radio wind soundings). They radio back information about temperature, humidity, air pressure, and winds. Each launch of a rawinsondes costs about $100. This may not seem like that much, except the $100 doesn't reflect the initial cost of setting up a receiving station on the ground, nor the cost of a tracking device to follow the balloon so winds can be computed. Such equipment costs tens of thousands of dollars.

Further costs are the manpower to operate the equipment for each launch, plus the time needed to compute and transmit the data. Thus, the daily operation of an upper-air station is costly, especially for poorer countries. In those countries, when a piece of equipment breaks down, it is sometimes months before a part can be obtained to correct the problem—months with no data being obtained.

The real problem with global coverage is that most people live on land, whereas 70% of the earth is covered by water. Scientists need data obtained over the water in order to make accurate weather forecasts that apply to the land. For example, the air above the eastern United States on any particular day very likely was over the ocean just two or three days earlier. No wonder the families who had made plans for their vacation based on a long-range forecast are disappointed when the "beautiful" weekend turns soggy.

FORECASTING AS ART

Watching weather is anything but dull. Sure, there are less eventful days, when a high-pressure area overhead offers nothing more exciting than a few fair-weather clouds. But the weather is always changing, even when it looks as if nothing is happening. Weather forecasters look forward to storms the way professional basketball players look forward to the championship series.

Snowstorms, for example, offer a chance for weather forecasters to practice their art. How can one predict whether or not the precipitation in an approaching storm will occur as rain or snow? Predicting the rain/snow line is the artistic part of weather forecasting. Sometimes a shift of the storm center by a mere 50 miles can mean the difference between a foot of snow and a rainstorm. Because the data from upper-air measurements are recorded at intervals greater than 50 miles horizontally, the weather forecaster is often forced to apply art where science has left him or her lacking.

Modern weather forecasts, the kind you see on television and hear on the radio, are generated by computer models that use data obtained primarily from the upper-air network. Other data derived from our current satellite capabilities are also used by the computer models. Whereas all balloons are launched at the same time—noon and midnight, Greenwich Mean Time (GMT)—the satellite information has to be "assimilated" so that it looks as if these measurements were also taken at noon and midnight GMT.

Forecasters use the term progs as shorthand for prognoses, the forecasts that result from the data entered into the computer model. These progs are used by the National Weather Service and such private forecasting companies as Accu-Weather and the Weather Channel to make forecasts. These progs are the "science" part of the forecast. Such forecasts are distributed by the National Weather Service to their forecasting offices throughout the United States, where they are further distributed to the news media, private forecasters, and the public. Although these progs have been tailored for the specific geographical area, they are based on computer models and satellite information gathered from a much larger geographical area.

Suppose a TV forecaster gets one of these forecasts on his wire in Pittsburgh while planning his evening weather forecast for the metropolitan area. According to the forecast based on the computer model, the city should be receiving heavy rain with temperatures in the upper 30s that evening. But 80 miles to the west, in Youngstown, Ohio, the weather forecaster there is predicting a foot of snow for that city, based on the same NWS forecast.

The forecaster in Pittsburgh thus realizes that the rain/snow line is somewhere between Youngstown and Pittsburgh, a distance of about 80 miles. But the data that went into the forecast model—the data recorded on the most recent upper-air balloon launch—was gathered at sites more than 200 miles apart. So between Pittsburgh and Youngstown there is a great deal of atmosphere that hasn't been observed, atmosphere that represents blind spots in our knowledge and thus weakens our ability to make accurate forecasts.

Compounding the difficulty is the very nature of atmospheric processes. They are highly nonlinear, meaning that abrupt changes can, and do, occur within very short time and space scales. Most of us have experienced sharp changes in the atmosphere—the sudden cessation of a strong wind, the abrupt drop in temperature after a wind shift.

Sometimes such nonlinear changes take on a bizarre twist, forcing us once again to feel humble before the primal force of nature. While living in Boulder, Colorado, one of the authors was forced to cancel a dinner invitation in Denver, 30 miles to the southeast, because of a snowstorm that dumped a foot of snow on Boulder and caused his car to slide into a ditch after a shopping trip. When he called to explain why he would not be able to attend the dinner that evening, the host was skeptical, since the skies over Denver were clear and the weather forecast was not calling for any precipitation in the surrounding area.

Adding to the host's skepticism was the fact that in Ft. Collins, 40 miles north of Boulder and the site of the closest NWS forecasting office north of Denver, there was no mention of snow either. But that's the way nonlinearity is manifested in the atmosphere. If the forces that create weather were linear, then you could get a fair reading of the weather in Boulder by averaging the weather in Denver and Ft. Collins.

Such an example—the sudden, brief genesis of a small-scale, local snowstorm—illustrates the failure of the existing upper-air network to provide enough information to predict such an occurrence.

If you are the weather forecaster in Pittsburgh, and the NWS chart tells you the city can expect heavy rain while 80 miles away the residents can expect snow, what can you do? The first thing is to realize that the chart's prognosis was obtained using data gathered at 200-mile intervals, so when it comes to fine-tuning a forecast the forecaster must rely on his own resources.

Many local forecasters have a network of ground observers to supply them with information more localized than that provided by the NWS charts. With these ground observers, a simple series of telephone calls can determine, for example, exactly where the snow/rain changeover line is. This information, coupled with the forecaster's own experience of the local geography and microclimate, can result in a much more accurate forecast than one that exclusively relies on computer models.

For example, our forecaster in Pittsburgh might gather information from his ground observers in the small towns and suburbs to the west. He might discover that the wind in that vicinity is coming from the northeast, and he knows that in that particular geographical area of Pennsylvania and Ohio a northeast wind usually means snow. Experience and intuition fill in the gaps left by science. Again, we are talking about the "artistry" of weather forecasting.

But still, basing a weather forecast on ground observations is a risky business. What would be ideal is knowledge of the vertical structure of the atmosphere based on observations made at closer intervals. Such observations would result in a weather chart with finer resolution. It is exactly this kind of measurement that will be available some day in the near future, for that's when a series of new satellites will be launched that will make our present methods of gathering data seem as obsolete as using an abacus to balance a checkbook.

These new satellites will be of two kinds. One set will be of the geostationary variety. Geostationary satellites orbit high above the earth at about 22,300 miles. They provide the pictures of the earth we see on the television weather forecast, the so-called "satellite shot." As they scan across their field of view from their lofty perch, these satellites can send back information about a particular region of the earth every half hour. Their disadvantage is that because of their altitude, the resolution of such pictures, that is, the detail with which they can make measurements, is limited.

The first of a series of new geostationary satellites was launched in 1994 equipped with more sensitive instruments to supply a wider range of data at a higher resolution to ground stations. Only now are meteorologists getting used to the vast amounts of new information available to them. For the most part, they are just beginning to use all of their new information to improve their forecasting capability.

Another set of satellites will be launched later, probably early in the next century. These satellites will be polar-orbiting platforms (i.e., their orbit takes them around the earth on a north-south longitudinal course) that will be capable of probing the earth's atmosphere with lasers. Such probes should be able to furnish measurements of the atmosphere with a spatial resolution heretofore thought to be impossible. Polar-orbiting satellites circle the earth longitudinally rather than latitudinally about once every 90 minutes, at an altitude of 500 miles. One polar-orbiting satellite can provide one measurement of the entire earth once a day. A series of such satellites, operating simultaneously, can provide complete measurement of the earth's atmosphere two, four, or six times a day.

Such an influx of information will require a whole new generation of computers able to input such data and produce models. Figure 2 shows the kind of resolution that the Laser Atmospheric Wind Sounder (LAWS) can obtain. Using

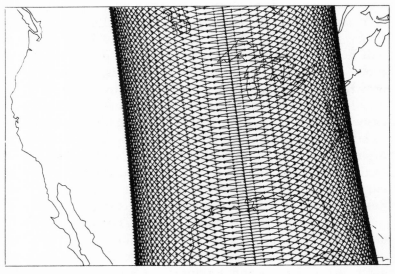

FIGURE 2. An example of the scan pattern for a polar-orbiting, space-based Doppler radar. Some information about the winds is obtained continuously along these scan paths and detailed information about the vertical distribution of the winds could be obtained at every location where the lines of the scan pattern intersect. Thus, wind data for use in weather forecasts could be input at a resolution of tens of miles, rather than hundreds of miles, as is currently the case. (From Curran et al., 1987.)[4]

this instrument, a weather forecaster would have data throughout the entire atmosphere obtained from readings taken every 2 miles, rather than every 200 miles, as is currently the case. Similar laser instruments are being planned that will measure water vapor.

With such new instruments and satellites, weather forecasters in Pittsburgh, Des Moines, and San Francisco will have a much better picture of their local area and won't have to resort to intuition and art to provide an accurate forecast. With such instruments, the snowstorm that hit in Boulder wouldn't have been a surprise. Forecasters will have ample information to make an accurate forecast well in advance.

We've already made significant progress in weather forecasting—even since the blizzard of 1978. Thanks to satellites seeing every corner of the globe, the death toll from a series of storms crashing into Southeast Asia in 1992 was kept down to a few thousand whereas the same kinds of storms hitting that region would have killed hundreds of thousands of the unprepared populace only a decade earlier. The science of weather forecasting has advanced to such a state that we think we even know how far ahead a truly accurate forecast can be valid. Professor Edward Lorenz, of the Massachussetts Institute of Technology, has utilized the complex mathematical tools of chaos theory to conclude that the equations used to describe atmospheric motions can be solved accurately for no more than ten days to two weeks, regardless of how detailed and accurate the data going into the equations are.

Nonetheless, a revolution in weather forecasting is just about upon us. These new instruments and the computers that will process their information will also provide us with new insights into the forces that drive the weather. These forces also determine the earth's climate, so that our understanding of long-term weather patterns and climatic trends will also deepen and increase.

Imagine some day in the near future. It is July and you are planning a vacation in Florida in October. You hear a long-range forecast that calls for a hurricane to hit the Florida coast on September 2. Absurd? According to Lorenz's chaos theory, such a prediction is only fiction. Then again, maybe a new generation of mathematics may evolve that will make chaos theory as out-of-date as the most sophisticated slide rule.

TWO

Red Sky at Night: The Scientific Basis for the Old Proverbs

There was a time before television weather forecasters. When the first issue of *The Old Farmer's Almanac* was published, there were no radios, no televisions, and only a smattering of newspapers in the newly independent "colonies" of England.

How did people back then know when to plant corn, harvest pumpkin, or sow wheat? How did they know whether to wear their light muslin shirt or to carry along their heavy mackinaw? They didn't have the news and weather on the hour to listen to, nor could they simply turn on the National Oceanic and Atmospheric Agency (NOAA) 24-hour forecast.

They did have their eyes, ears, and other senses. They could look out the window, maybe go outside and wet their thumb and hold it up to tell which way the wind was blowing, bend their rusty joints to see if the arthritis was acting up. These were all clues, clues to the atmosphere, clues to what kind of weather was on the way.

To aid them in this task, the old-timers had a store of knowledge passed on from generation to generation. These were the old weather proverbs, short sayings, often rhymes, that contained useful information. Some of these verses are

Red sky at night, sailor's delight; red sky at morning, sailors take warning.

When leaves show their undersides, be very sure that rain betides.

Curls that kink and cords that bind; signs of rain and heavy wind.

Sound traveling far and wide, a stormy day will betide.

These simple rhymes, old proverbs that have been passed down for centuries, were based on observation. Although simplistic, they point out some very basic general rules about weather dynamics. To understand these simple proverbs, and to understand the coming revolution in weather forecasting, it is useful to explore the basic fundamentals of weather, the forces that shape our day-to-day environment.

SIMPLIFIED WEATHER

All weather is a simple transfer of energy. The source of our weather on earth is the sun; heat and light constantly radiate from the sun and are intercepted by the earth. The astronomer Edmund Halley is credited with establishing the relationship between the earth's winds and the energy from the sun. Halley attempted to describe the findings of the explorers who were the first Europeans to observe the torrential rainfalls of the monsoons and the prevalence of winds blowing from east to west—quite unlike the wind patterns observed in their homelands. He noted that these recently discovered features of the weather were present in the Atlantic, Indian, and Pacific Oceans, and therefore he concluded that some powerful external force was responsible for such a universal feature. In 1686, he wrote a detailed treatise implicating the sun as that source of energy.[5] The energy that creates life begins with the sun, and our weather is the manifestation of this energy—the winds, the swells of the sea, the rain and snow, the violent storms, all of these are generated by the sun's energy moving from one place to another.

If the earth did not move, but hung suspended in space, our weather would be simple and uncomplicated. The sun shines down on the earth most directly at the equator, most obliquely at the poles. If you shine a flashlight on a wall at a right angle, you will notice that the ray of light appears quite intense. If you change the angle, the intensity diminishes; the more oblique the angle, the less intense the light beam. This relation holds for the sun and the earth. At the equator the sun's effect is more direct, the heat and light are more intense. At the poles, the shadows are longer, the light hitting the ground at a steeper angle, and thus the heat from the sun is less.

Another of the early general observations about the atmosphere noted by the sailors to the new lands in the 17th and 18th centuries was that winds *generally* carried air from east to west (easterlies) at low latitudes but that they *generally* came out of the west at middle latitudes (westerlies). The global aspect of these circulation characteristics is called the *general circulation* of the atmosphere. George Hadley wrote the classic paper in 1735 that linked rising motion in the tropics and the rotation of the earth to the existence of easterlies at low latitudes. He further noted that the return flow must then be at higher latitudes and that this descending air was the primary cause for the general feature of the westerly winds at middle latitudes. This feature of ascending air at the equator and descending air at higher latitudes is an important indicator of the general circulation and is referred to today as the Hadley cell.

Many modifications have been made to Hadley's original model and we now know that Hadley's simplistic view violated some of the basic laws of physics. It is known that the downward branch of the Hadley cell does not extend to middle latitudes, but marks the boundary between tropical regimes and temperate climates more characteristic of the middle latitudes. The location of the downward branch of the Hadley Cell is in the region that sailors came to know as the doldrums, latitudes generally between 20 and 30° where there was very little wind. Because descending air loses its moisture, the dominant downward motion in these regions over land creates some of the great deserts of the world.

The weather at middle and high latitudes, on the other hand, is dominated by the movement and interaction of very different air masses, a phenomenon first described by a group of Norwegian scientists in the late 19th century. It wasn't until the middle of that century that scientists realized the shortcomings of Hadley's description of the general circulation as they tried to explain Hadley's motions using mathematical formulas. What they discovered was that the weather in the tropics was decoupled from the weather at middle and high latitudes, although Hadley's model was credible for describing the general circulation of the tropics. Even now, as scientists are able to observe the entire planet using satellite-borne sensors, they are learning even more about the interaction between the tropics and middle latitudes.

THE MOVEMENT OF WEATHER

The basic "weather machine" of the earth is that of the sun heating the equator, the hot air rising, releasing its moisture in the form of rain and thundershowers, then spreading north and south and descending as cooler and drier air. Another significant aspect of the atmosphere is that it holds moisture in the form of water vapor, which is simply water droplets that have been transformed from a liquid into a gas. In the process of rising, the moisture in the hot air condenses back into liquid as the air rises and cools. This process of convection creates clouds and rain. Near the equator, convection produces a thundershower almost every day. The same air that circulates downward after rising and cooling has little moisture, and what moisture is left soon dissipates as cool air descends.

But the variability of our weather, especially in the temperate latitudes between the tropics and the poles, is caused by a variety of factors that influence the movement of the sun's energy to other parts of the globe. The earth, after all, does not hang suspended in space, motionless. It is moving all the time. It rotates on its axis, it revolves around the sun, it tilts toward and away from the sun at different times. All these movements affect our weather.

We have noted that the sun's rays land more directly on the equator than on the poles. Thus, the equatorial region, the "tropics," is the furnace of our weather, the area where the sun's energy has the most direct effect. Knowing that, it is easy to see why the temperature varies over the rest of the planet. The sun's rays do not land as directly on the surface, and therefore they contain less energy.

The differences in air temperature over the earth's surface are what cause winds. From the global perspective, winds are the means by which the energy produced at the equator is disseminated across the globe. Generally, air over the equator is warmer than air over the poles. Warm air is lighter than cold air and thus tends to rise. Such convection creates a vacuum which nearby air rushes to fill, thus creating wind.

On a smaller scale, this is obvious if you go to the beach in the morning on a sunny day. When you arrive the breeze may be blowing from land out to sea, because in the morning the air over the water will be warmer than that over the land, since the land cools faster than does water. As the sun heats up the land, by noon or thereafter, you will notice a shift, to a cooling breeze coming in from the ocean, known as a "sea breeze." This occurs because the air over the water is now cooler than that over the land. The heated air rises and the cool air over the sea moves landward to fill the void. This land-sea breeze circulation is driven by the same mechanism Hadley observed in the 18th century. But the small-scale direct circulation is not affected by the large-scale rotation of the earth and what we observe is a simple two-dimensional phenomenon: onshore in the heat of the day, and offshore the rest of the time.

THE HIGHS AND LOWS OF WEATHER

If the earth didn't rotate, this movement of air would be quite simple to predict. The United States would experience mostly cold temperatures as the air from the poles traveled down our continent to the tropics. But since the earth does rotate, this movement causes the air to twist into spirals. We know these spirals as high- and low-pressure areas, the same air masses we see on the weather map every evening on television as the weatherman "predicts" our weather.

This force of movement, caused by the rotation of the earth, is called the *Coriolis force*. Imagine being on a merry-go-round with a partner. Both of you are sitting on horses at opposite sides of the platform. If the merry-go-round were not moving, and if there were no posts in the center, one of you could easily throw a ball straight across the center of the circle to the other. But if the merry-go-round started up and began spinning, you would find that the same thrown ball would now twist forcefully. This is what the spinning earth does to air currents.

Again, if the earth weren't rotating on its axis, there would be no Coriolis force. In such a case, differences in air pressure would appear like stationary ridges and valleys. This is true because our atmosphere has weight. Although it's difficult to conceive of, some air is heavier than other air, that is, more dense. Air that is denser exerts more pressure on the surface. Although we can't feel this pressure, we can measure it with a barometer. When the air pressure is high, it means that the denser air has piled above our geographical area. This "piling on" of dense air forms a dome, or ridge. This is why television weathercasters often speak of a "ridge of high pressure."

Air that is less dense exerts less pressure on the surface, thus forming a depression, or valley, rather than a ridge. Overall, then, the atmosphere is quite uneven, with ridges and valleys representing the high- and low-pressure cells. Cold air from the top of a high-pressure ridge flows down toward the depression at the center of a low-pressure cell.

Since the earth rotates, however, the Coriolis force causes winds in the northern hemisphere to spin clockwise away from the center of high-pressure areas and counterclockwise toward the center of low-pressure depressions. In the southern hemisphere the directions are reversed. Air in high-pressure areas travels counterclockwise, air in low-pressure areas spirals clockwise, as a result of the same Coriolis force.

These high- and low-pressure areas are characterized by distinctive weather. In a low-pressure area, the air on the ground is moving inward, toward the center of the depression. When it gets there it rises quickly. This uplifting of the air causes the moisture in the air to condense, creating clouds and precipitation. Thus, low-pressure areas are usually associated with inclement weather.

In a high-pressure area the air at the ground is moving outward and downward, sliding down the dome of high pressure. This downward movement causes the air to warm, thus drying it. High-pressure areas are usually associated with clear, dry weather.

Why is this so? The answer is that cool air holds less water vapor than does warm air. Thus, when warm air is cooled, the water vapor in it must condense into moisture droplets, that is, rain. Water vapor is a gas, invisible like air. Yet we can feel it. On a hot summer's day, we often describe the effect of water vapor as "muggy," "humid," and the like. This is because hot air is able to "hold" more water vapor than can cooler air.

If you take this warm air, full of water vapor, and cool it quickly, the water vapor is "squeezed" out of it by condensation into water droplets. Therefore, precipitation occurs when warm air, full of water vapor, is cooled. This occurs most often when warm air rises to higher altitudes, because pressure in the atmosphere decreases with altitude and a decrease in air pressure causes the temperature to drop.

The average pressure exerted by air at sea level is about 14.7 pounds per square inch, although you'll never hear atmospheric pressure expressed in these units. What you will hear is inches of mercury (sometimes millimeters of mercury), millibars, and hectopascals. Inches of mercury really isn't a measure of pressure; it refers to how high a column of mercury is pushed by the atmosphere. The first barometers were long (about 3 feet) tubes filled with mercury. When these tubes were turned upside down and allowed to sit in a bowl of mercury, the pressure from the atmosphere pushed the mercury up the tube about 30 inches. As the air pressure increased or decreased, the force exerted on this bowl of mercury caused the mercury to rise or fall in the tube. What we commonly hear over the radio or television is how high the column of mercury has risen in the tube. In actuality, any fluid can be used, but mercury is used most frequently because it is so dense. If water were used, for example, the tube would have to be more than 30 feet long.

One of the correct units of pressure that is traditionally used by meteorologists is the *millibar* (abbreviated mb or mbar). As the general public becomes more educated about the weather (thanks, in part to the success of the Weather Channel) they will become more familiar with the term millibar. This is the unit of pressure most meteorologists have been taught to use. Normal atmospheric pressure at the surface is about 1000 millibars.

Meteorologists, however, are the Rodney Dangerfields of science. They get no respect from the more formal scientific disciplines such as physics and chemistry. Colleagues in the latter sciences have criticized the use of the millibar as a measurement unit. The scientific community has mounted a concerted effort seeking to require that a particular physical property (such as pressure, speed, or density, etc.) be characterized using the same units regardless of which scientific field is doing the measuring. In this way, the same units of measure would be heard in a conversation among, say, a meteorologist, a physicist, a chemist, and a geologist.

Imagine the confusion that results when a meteorologist talks about pressure measured in millibars to a chemist who measures pressure using the torr unit. Meteorologists being educated today use the hectopascal (hPa) unit which is interchangeable with the millibar and which the physicist is already familiar with and which the chemist is also learning to use. Thus, some day, you may hear the word "hectopascal," describing the pressure.

It may seem odd to talk about air weighing anything at all, but it does. In fact, it is the weight of the air exerting pressure on us that keeps our bodies intact. If we were suddenly catapulted into outer space, away from our atmosphere, our bodies would explode, shouting "free at last." That's why astronauts wear pressurized space suits.

As air rises there is less air above it and less pressure; therefore, according to one of the laws of physics, the temperature decreases. This does not mean that low pressure areas have cold temperatures and high pressure areas have warm temperatures. In fact, high or low pressure implies nothing about temperature. It is the *process* of increasing or decreasing pressure that causes the *relative* temperature to change according to altitude.

As a general rule we can say that clouds and precipitation are formed by rising air (decreasing pressure causes warm air to cool, which in turn condenses water vapor), and clear weather is produced by sinking air (increasing pressure causes cooler air to warm, thus drying it out). This is why the tropics are the home of our rain forests, where showers fall every day as a result of the rising moisture-laden air condensing into thunderclouds. And it is also why the sinking air over the Sahara desert is devoid of moisture when it reaches ground level.

This then is the basic machinery of weather—the movement of heat, the dissemination of the sun's energy throughout the atmosphere. Weather is the consequence of various conditions that create change, for example, when there is too much heat or too little moisture. Seeing the weather machine as a thermostat is helpful in understanding the "why" of weather. A thunderstorm is a means of transferring energy. When there is too much heat caused by the direct sun, a thunderstorm cools things down.

This is what makes weather so exciting; it is always changing, always dynamic. At each moment, some part of the earth is absorbing energy from the sun. And this energy is somehow distributed to other areas of the planet. If you think of the earth as a living organism, then weather can be regarded as the equivalent of the circulatory system of the human body, bringing nutrients to the far-flung corners of the organism. Weather, too, brings nutrients, in the form of heat and moisture, to areas outside those that benefit from the sun's rays most directly.

COMPLEXITIES OF WEATHER

So far we have discussed the components of weather in fairly simple terms, as if the earth were a static organism subject to the laws of nature in a uniform manner. If that were the case, forecasting the weather would be an easy matter. But we know that weather is dynamic. It is this changeability that creates complexity, that defies our attempts to make *laws* to which the weather must adhere. So how are we to make accurate weather predictions? The answer is by studying the variables, the factors that influence the daily interchange of the earth's energy. By looking upward, by studying the clouds, sky, and precipitation, we can, like our ancestors, forge some progress in understanding what creates our weather.

Clouds

Since our weather comes from the sky, that is where we must look for a clue to the coming day. In ancient times, before the invention of the barometer, a look at the sky was about all that our ancestors could employ to tell them whether the next day would be a good one to spend harvesting or hunting or a day best suited to finding shelter.

A quick look at the sky can tell you a good deal about what kind of weather to expect. While a fair sky almost always means good weather, a cloudy sky doesn't necessarily mean that precipitation is imminent. The shape, density, color, and variety of clouds in the sky determine the type of weather you can expect.

As we've noted, clouds form when enough water droplets mass together. Clouds are composed of millions upon millions of tiny droplets or ice crystals. Although clouds appear to have substance, i.e., to be solid, they are about as solid as fog. Fog, as most people know, is a cloud that has formed at the earth's surface. So if you've ever walked through fog you've walked through a cloud.

Clouds appear solid because from the ground they look

white. But the color of clouds is merely the reflection of sunlight. During most of the day they are white, but at sunset they reflect the colors of the rainbow.

In the early 19th century scientists noticed that clouds appeared to assume three basic patterns: curly or fibrous, layered or stratified, and lumpy or heaped, which they termed *cirrus*, *stratus*, and *cumulus*, respectively.

Further, it was noted that clouds form at different heights in the atmosphere, and on this basis we can divide them into two groups: those whose bases usually lie more than 6 kilometers (km) (3.7 miles) above ground, which we label with the prefix *cirro*; and those whose bases lie between 2 and 6 km (1.2 and 3.7 miles), which we label with the prefix *alto*.

We can use these five combinations to describe the pattern and the height of the clouds in the sky. But what do the shapes, textures, and heights of the clouds tell us? Plenty. Let's look at the lumpy clouds first, the *cumulus*.

A cumulus cloud resembles a cauliflower. Cumulus are the small, "cotton-puff" clouds you can see on a fair day in the afternoon. They are formed by rapidly rising hot air currents, called thermals. On sunny afternoons at the oceanfront, the sky over the land will be dotted with cotton-puff clouds, and yet over the water the sky is cloudless. That's because the air over the ocean is cooler than that over the land, and thus no thermals rise to form clouds.

Generally, cumulus clouds are associated with fair weather. But sometimes, on very hot, humid days, or when there is a rapid uplift of warm air as occurs when a cold front is approaching, the cute little "cotton-puffs" grow. And grow. When they become quite large and they extend to 30,000 or 40,000 feet, their appearance changes. The entire top may rise to the level of the jet stream where the strong winds at that altitude flatten them out, producing an anvil shape. The sky darkens as the massive cloud, rising up and spreading out, covers the entire sky. The mild-mannered cumulus has been transformed into the *cumulonimbus*, the fiery harbinger of thunder and lightning that we associate with summer downpours and tropical monsoons.

Cumulus clouds that form when a thermal layer prevents the rising thermal from ascending higher may exhibit flattened bottoms. These are called *stratocumulus* if they form at a lower level, and *altocumulus* if they form at a higher level. All of the cumulus clouds are usually indicative of fair weather, as long as they don't grow too large and become threatening *cumulonimbus* clouds.

Cumulus clouds are easy to identify; they sit in the sky with their lofty peaks like little creampuffs, well defined. But *stratus* clouds are different. They are situated lower in the atmosphere than cumulus, and often cover the entire sky. Stratus are the clouds that cause the "gloomy" general overcast of a steady rain associated with a warm front.

What distinguishes stratus clouds is their very lack of distinction. From the ground they have no characteristics. The sky is simply gray, with no distinguishing features to the cloud cover. Stratus clouds are formed by the general lifting of an entire layer of air, such as occurs when an advancing warm front is lifted over the cold air before it, condensing the moisture into large stratus clouds that cover the sky and produce a steady rain.

As the moisture-laden air is "wrung" of its moisture, the stratus clouds change as they rise in the atmosphere, becoming *altostratus* and, even higher, *cirrostratus*. Cirrostratus clouds are the high, thin clouds that produce a "halo" around the sun or moon, and have been associated with an advancing warm front that will bring rain within 24 hours.

The highest clouds in the atmosphere are *cirrus*. These are the delicately laced fibrous "tails" seen high in the sky on clear days. Cirrus clouds are composed entirely of ice crystals, and their long, thin shape is the result of the ice crystals being blown by upper-level winds.

PRECIPITATION

Rain and snow are products of clouds. Although not every cloud produces precipitation, no precipitation can be produced without clouds. Clouds are produced by rising air, air full of water vapor that condenses into water droplets as it is cooled by altitude. But this is just the beginning, as condensed water vapor is not heavy enough to fall from the cloud as precipitation.

How do raindrops grow? The first step is called nucleation. Water vapor needs something to grow on, some tiny island on which to anchor itself in the midst of the air molecules. Dust and microscopic pieces of sea salt serve this purpose very well. In fact, if the atmosphere had hardly any dust and dirt in it, we'd see a lot less precipitation. Dirty, dust-laden air produces precipitation quicker and in greater volume than clean air. This occurs because the foreign substance—the molecule of dust, or some other gas such as sulfur dioxide breaks down the surface tension on the water molecule and allows it to combine with the foreign substance to form a water droplet. Droplets then grow, collecting greater mass until they eventually become heavy enough to fall out of the cloud. With enough water droplets, you have rain.

Clouds that form over land can contain up to 1 billion droplets per cubic meter. Those forming over the ocean contain less. These water droplets, on their way down, pick up other droplets in a process called coalescence. If there are enough water droplets, the droplets end up striking the ground as rain or snow. If there are not enough droplets, or if the air below is very dry, the precipitation will often evaporate before reaching the ground, a phenomenon called *virga*.

Whether the final result is rain, snow, or hailstones, the beginning of the process was precipitation in the form of water droplets. Whether these droplets end up on the ground as rain or snow depends on the temperature of the air through which they have to travel. Normally, air temperature decreases with increasing altitude. It is colder in the clouds than it is on the ground.

But sometimes the temperature, where the clouds are making precipitation, may actually be warmer than the air temperature on the ground. When that occurs, the precipitation leaving the clouds is in the form of rain, but if the temperature near the ground is below the freezing point of water, we have *freezing rain*. If that same raindrop freezes on the way down, then we have *sleet*.

SKY COLORS

All of this brings us back to our proverbs. Is it true that a red sky at night predicts a clear day tomorrow? Or that a red sky in the morning means rain is on the way? To answer these questions on a scientific basis, we need to know the answer to one of life's basic questions: why is the sky blue?

The light we see radiating from the sun is but a small part of the spectrum of energy that is our lifeline. Visible light is only one type of electromagnetic radiation emanating from the sun. Radio waves, infrared, ultraviolet, and microwaves, are some other components of the sun's magnificence. But let's concern ourselves here with the visible band of light, because we want to know why when we look up at a clear sky we see the color blue.

Light, we now know, travels as both wave and particle. The length of the wave determines what color our eye sees. At one end of the spectrum is the ultraviolet, the shortest wavelength. At the other end is the infrared, the longest wavelength. In between these two invisible (to the human eye) wavelengths lie the colors of the rainbow—blue, green, yellow, orange, and red.

We know that the earth's atmosphere is composed of matter—water molecules, dust, atoms of oxygen, nitrogen, and other gases in such small quantities they are called "trace gases." When sunlight approaches the earth, it has to travel through the layers of the atmosphere to reach us. In the process, certain wavelengths of the light spectrum are absorbed while others are refracted by the dust and matter in the atmosphere. That is why the sky is generally blue—the rest of the spectrum has been absorbed by the atmosphere, and only the blue wavelength remains.

When the sun is visible, the sky appears blue because the light rays are entering our atmosphere directly. However, when the sun is close to or beneath the horizon, its rays shine obliquely through the atmosphere, and continue on to space. This means that the length of the absorbed wave changes, and instead of a blue sky we often see a red sky, or to be more precise we see a sky that is bluer away from where the sun's rays are cutting through the atmosphere and a red of various hues in that part of the sky where the sun's rays are manifesting.

This occurs generally at sunset and sunrise, when the sun's rays are longest. The redness of the sky at such times is caused by the refraction of the sun's rays by dust and water vapor. So, if at sunset the sky is red, it means that to the west there is dust in the air, a sign that the air is dry. Since in the northern hemisphere our weather comes from the west, a red sky at night would generally mean a fair day coming up.

A red sky in the morning is a bit more difficult to explain. Depending on the shade of red, and what else appears in the sky, a red sky in the morning could foretell a fair day or a wet one. If the red is somber, and quickly changes to gray as the sunlight advances on it, then there is moisture in the air and rain can be expected. But sometimes the red in the morning is simply the refraction from dust particles, the same as at sunset, and therefore a dry day can be expected.

On the other hand, a gray, gloomy morning usually foretells a gray, rainy day, right? Not always. If the gray and gloom are caused by fog, then we can expect a fair day. Why? A look at what causes fog provides a ready answer.

Remember that the warmer the air, the more water vapor it can hold. During the day, the air heats up over land. At night this same air cools. The heat absorbed by the ground radiates back to space on a clear night, cooling the surface even more. As warm air drifts in to take the place of the hot air rising on radiation currents, the water vapor condenses as the air is cooled. This produces fog.

When the sun comes up, after a night of radiation cooling, the ground is covered with fog. But several tens of meters above the ground, the sky is clear. After a couple of hours of sunlight to burn off the fog, the day does become fair.

So it is true that *a foggy morning does a fair day betide*. But "often" isn't good enough if your life depends on the weather and an accurate forecast. With the growth of scientific instruments and knowledge, weather forecasting has moved far beyond the proverbial stage. We are on the doorstep of a new age in weather forecasting accuracy.

THREE

Weather Forecasting and the Computer

A SHORT AND INCOMPLETE HISTORY OF COUNTING BEFORE THE COMPUTER

History doesn't tell us exactly when it was that human beings learned to count. But we do know that it was a long time ago—before electricity, before the emergence of nations, way back in primitive time. Perhaps it was a shepherd who decided that, to keep track of his flock, he would take a pebble and put it in his bag for each sheep that left the pen in the morning, and take a pebble out of his bag each evening as the sheep returned. If any pebbles were left over, he would know he had lost a sheep.

We do know that as early as the 11th century British law required its witnesses, to be credible, to be able to count to nine. But counting to nine or ten was easy, all we had to do was count our fingers or toes. What about counting larger numbers? What about a system of "remembering" which numbers had already been counted? Interestingly, there are records of prehistoric drawings in caves that may

41

represent our first attempt at numerical notation. Numerical notation was necessary when dealing with any numbers larger than ten.

We know that the ancient civilizations of Greece and Rome devised means of additive notational systems, in which one distinct symbol represents each different unit in the number base. The Roman numeral system in use today is an example of this early form of additive notational system. The Roman system uses capital letters, such as X, C, M to represent numerical values. Thus, a Roman could write the number 3745 as MMMDCCXXXXV, in which M = 1000, D = 500, C = 100, X = 10, and I = 1. It was quite adequate for dealing with numbers under 5000, but it became rather burdensome when doing mathematical calculations.

As civilizations grew and people became more curious about their world, it became necessary to figure out ways in which to calculate numbers. One method was what we call today positional number systems, in which the values being represented are denoted entirely by the *position* of the symbol (number) within the string of symbols representing the numbers. This may seem evident to us now, because it's the way we were taught math, but at the time it was a major breakthrough in problem solving. With a positional number system, the symbol 2, for example, can have a value of two or two billion, depending on its place in a string of symbols that comprise the entire number: 2 versus 2,000,000,000.

With a positional number system, civilizations were able to thrive, mathematically. Using this system, it was possible to calculate higher numbers quickly. At first traders would use pebbles and lines drawn in the dust to set up their positional number systems. Then as language became more sophisticated and the invention of paper made writing possible, such things as counting boards and abaci were developed.

Historians believe that the abacus originated in Europe and later appeared in the Far East, where it continues to be used today. Historians speculate that the abacus was developed as a means of making portable a system of using pebbles and lines drawn on the ground. The word itself is thought to be derived from the Semitic word *abaq*, which means dust, and refers to the practice of using pebbles and lines drawn on the ground. The abacus as we know it today is actually a result of evolution from the European abacus to the Oriental abacus, using wire and beads instead of pebbles and lines drawn on the ground or on boards.

The Oriental abacus was developed in the Middle East during the Middle Ages, entering the Orient via the usual trade routes. The abacus consists of an oblong frame of wood with a bar running across its length, dividing the frame into two horizontal compartments. Through this bar, at right angles to it, are placed small dowels, each strung with seven beads. Usually there are seven dowels, but sometimes there are more. Two beads are strung on the upper side, above the horizontal bar, and five are strung below the bar. Each bead in the lower section is worth one unit, while those in the upper section are worth five.

The Japanese revised the Chinese abacus by replacing the two beads on the top with one, and using only four beads on the bottom. This instrument is called a *soroban*. A skilled soroban operator can make calculations faster than someone working with pen and paper.

The abacus and soroban were only two of many attempts made throughout history to speed up the ability to calculate numbers. But calculating numbers was still time-consuming and burdensome. By the 17th century, as trade and commerce grew, vast armies of bookkeepers were hunched over paper and quills calculating. Not only were trade and commerce growing during this time, but so was science. Mathematics, after all, is one of the fundamental tools of scientists. And as science progressed from the Middle Ages to modern times, mathematicians and scientists were scrambling to come up with some device that would make calculations easier and faster.

At first such devices were simple embellishments of the old pebbles and lines. One device, invented by John Napier, the inventor of logarithms, was called Napier's bones and consisted of a matrixlike grid with the multiplicand at the head of the columns and the multiplier at the end of the rows. By moving the rows and columns around, you could quickly find the product of any combination of multiplied numbers.

As technology grew, mathematicians began to experiment with mechanical devices to speed calculation. The first known mechanical adding machine was invented by the German Wilhelm Schickard (1592–1635), although historians have been unable to find any drawings or plans of the machine. Schickard's invention may have been ahead of its time, for it was the French mathematician Blaise Pascal (1623–1662) who is generally considered as the father of the mechanical calculating machine. Pascal's machine consisted of a box small enough to fit on top of a desk or small table. The upper surface of the box had a series of small windows, below which were a corresponding series of toothed wheels. One turned the wheels and saw the numbers in the window above it.

Although primitive by today's standards, Pascal's machines inspired innovators such as Gottfried Wilhelm Leibniz, Charles Babbage, and George Scheutz to make further developments in succeeding years. These men and others expanded the capabilities of such machines to include more sophisticated mathematical calculations. By the beginning of the 20th century, a new development allowed operators to perform massive tabulations undreamed of decades before. This was the punched-card machinery, a system using cards with holes punched in them to record and tabulate data.

The primary impetus behind the development of punched-card machinery was the United States Census Department. During the latter decades of the 19th century the population of the United States was growing so quickly that the old methods of census-taking were proving too slow and costly, unable to keep up with the growing numbers of people. The early censuses were simple affairs: census takers visited households and asked some questions of the head of each family. In 1790, the year of the first census, the entire population was a little under 4 million. By 1850 the population had increased by more than a factor of 5, and the method of census-taking had changed also.

By the time of the 1850 census the Census Bureau was using a tally method. Each questionnaire was examined and a mark on a tally sheet was placed for each fact to be tabulated. Then the marks were counted, or tallied, for each district. You can imagine the hours needed to tally each census questionnaire. In fact, the 1880 census took close to seven years to compile, so that by the time the results were published they were out of date. So the Census Bureau committed itself to finding a method of tabulating the census figures for the 1890 census that would not take more than a few years to complete.

An employee at the Census Bureau, Herman Hollerith, had been working on just such a mechanical device. Hollerith was an engineer, and he devised an electric tabulating machine that used punched cards to record the data. His method was first tried during the 1890 census. Census takers showed up with their tabulating cards and punched out holes in the appropriate places. The cards were then run through the tabulating machine and tallied. The results were quite amazing. The 1890 census was completed within two years, and a rough count of the country's population was available six weeks after the start of the census.

The success of the 1890 census soon led to other commercial uses for punched-card machines. Companies such as IBM, Burroughs, and Powers began manufacturing such machines and selling them to industries, from the Ford Motor Car Company to the growing electric utilities. The punched-card devices continued to be used through World War II.

FROM COUNTING TO COMPUTERS

While such machines were proving a boon to business and industry, the fact that they were limited to mere tabulations of numbers proved a shortcoming for the scientific community. Scientists needed more sophisticated machines, something that would help them make mathematical calculations and apply mathematical formulas to solving problems.

As early as the 1920s Columbia University had established a Computing Bureau, using IBM punched-card machines. A decade later IBM developed and installed a "Difference Tabulator" at the university. The machine was used almost exclusively by astronomers to aid them in calculating planetary positions and the like. But there was a problem with the existing punched-card machines: they were very limited in what they could do. Although they could tabulate numbers, they couldn't deal with negative numbers. And they couldn't actually calculate mathematical functions automatically—they had to be operated at every stage by someone pulling levers or manning switches.

But as IBM saw the possible uses of such a machine in the physical sciences, it set about converting its punched-card machine into a more sophisticated machine that could be used by scientists to perform their calculations. At about the same time, Bell Laboratories was developing its own machine. Both companies started around 1937, and by the time World War II began both had electromechanical machines that could do scientific calculations at a faster speed than was previously possible.

Once again, it was a major war that provided the impetus needed for the development of an electronic computing machine. In the early days of World War II, at the Ballistics Research Laboratory of the Aberdeen Proving Grounds in Maryland, scientists were busy devising a means of using some kind of electronic computing device to shorten the time needed to compute the trajectories of bombs and cannon shells. The machine they built was called the Electronic Numerical Integrator and Computer (ENIAC). The first model was completed before the war's end and was operating by December of 1945. It required 18,000 vacuum tubes, 70,000 resistors, 10,000 capacitors, and 6000 switches. It was truly a monster, measuring more than 100 feet long and 10 feet high.

With the development of ENIAC, scientists finally had a machine that would aid them in their mathematical calculations. Even this most primitive computer made it possible to do sophisticated calculations with remarkable speed.

COMPUTERS AND WEATHER FORECASTS

One of the fundamental tools of the physical scientist is mathematics. Mathematics is the means by which physicists and chemists can describe the motions and interactions of atoms and molecules. Biological sciences, such as zoology and botany do not, in general, try to use mathematical equations to describe motions and interactions at the most primary molecular level, although mathematics and statistics are often used to explain the dynamics of cells and plant materials. Social sciences such as psychology and sociology involve so many parameters that mathematical equations are not very useful to describe or predict the behavior of people or groups of people. These sciences generally rely on statistical relationships to suggest how certain populations will behave.

Meteorology is generally considered a physical science, but in one respect it is vastly different from other sciences such as physics and chemistry. In physics or chemistry, scientists can devise laboratory tests with tight discipline and controls to determine the behavior of atoms or molecules. But meteorology has no laboratory, except for the vast atmosphere outside. Thus, meteorologists are faced with a dilemma: where can they test their mathematical equations? They can't, except by watching the future as it unfolds in the atmosphere.

This dilemma creates an interesting paradox because, in principle, the basis of meteorology is nothing more than the movement of molecules, which should be fairly simple to describe according to the laws of physics. So if the equations exist to describe the movement of air, but if such movements cannot be verified in a laboratory, what is the best way to describe and even to predict what will happen in the atmosphere? This was the problem facing meteorologists before the development of the computer.

BEFORE THE COMPUTER: THE EARLY DAYS OF NUMERICAL WEATHER PREDICTION AND THE DREAM OF LEWIS RICHARDSON

In 1812, the French mathematician Pierre-Simon Laplace wrote that "complete knowledge of the mass, the position, and the velocity of a particle at any single instant would allow for the precise calculation of that particle at any time, be it in the past or in the future." This continues to be the basic premise behind the science of weather forecasting: the physical laws that control the atmosphere can be used to determine its future state.

During Laplace's time, scientists were already familiar with the classical laws of mechanics described so eloquently by Sir Isaac Newton in his treatise *Principia Mathematica*. But what Laplace and his contemporaries did not know at the time were the fundamental laws of thermodynamics, which are crucial for predicting the motions and temperatures of fluids such as air and water.

This void was filled by the seminal work of the German scientist Rudolf Clausius, in the middle of the 19th century. Clausius identified the conservation of energy as the first law of thermodynamics and then went on to formulate the second law of thermodynamics. "In the absence of external constraints," he wrote, "the net flow of heat between two bodies is from the warmer to the cooler one." By the end of the 19th century, the world had recognized the fundamental laws of classical physics, and the goal of accurately predicting outcomes by numerical calculations was closer to reality.

One of the first meteorologists to recognize that the general laws of physics could be applied to the atmosphere was the Norwegian meteorologist Vilhelm Bjerknes who wrote in 1904:

> If it is true, as every scientist believes, that subsequent atmospheric states develop from the preceding ones according to physical law, then it is apparent that the necessary and sufficient conditions for the rational solution of forecasting problems are the following:
>
> 1. A sufficiently accurate knowledge of the state of the atmosphere at the initial time.
>
> 2. A sufficiently accurate knowledge of the laws according to which one state of the atmosphere develops from another.[6]

Bjerknes came from the Bergen School, which was recognized then as the premier university for the study of weather forecasting. It was at Bergen that the "polar front theory" was first developed. The polar front theory recognizes the importance of distinct air masses in understanding how weather changes. According to Bjerknes' theory, there is likely to be instability where two air masses meet. We take this fact for granted now, but at the time Bjerknes' discovery provided a solid basis for forecasting the weather. With the polar front theory it was possible to predict the weather in the future by a study of existing conditions. By analyzing the weather every few hours and producing synoptic maps (i.e., maps depicting the weather for that particular time) and then by following the features on these maps over several time periods, the forecaster could extrapolate into the future.

Lewis F. Richardson used the polar front theory as the basis for his classic work, *Weather Prediction by Numerical Process*, published in 1922. In that book he used a technique that he felt would allow him to forecast the weather over central Europe based on the weather at locations within a 500-mile radius of the designated area. Unfortunately, his numerical technique failed in its first trial. Using his formula, he predicted a pressure drop of 145 millibars (~4 inches of mercury) within six hours. Such a massive drop in barometric pressure over such a period of time had never been known to occur. Even the strongest hurricanes rarely achieved pressure drops of more than 100 millibars during their development, which normally takes place over a period of several days.

Despite its failure, Richardson's work was important in one respect: He was the first to envision a roomful of people performing the mathematical calculations necessary to forecast the weather by solving the basic equations that explained the movement of air so precisely described by Bjerknes. In other words, Richardson envisioned a human computer, with each person assigned a specific mathematic calculation.

In many respects, Richardson's numerical model bears a close resemblance to the first operational model of the United States National Meteorological Center (NMC), developed more than 40 years later. What Richardson did was to put a "grid" over the area in which the forecast was to be made. In this grid, he alternated input data for pressure and for winds just as today's numerical forecasts utilize different sets of variables alternating on the computational grid. The resulting input data grid resembled the checkerboard in Figure 3.

One important aspect of a numerical weather forecasting model is the third dimension, composed of several layers in the vertical. We know that the atmosphere exists in layers, and that conditions change as one rises vertically from the earth's surface to the limits of the troposphere.

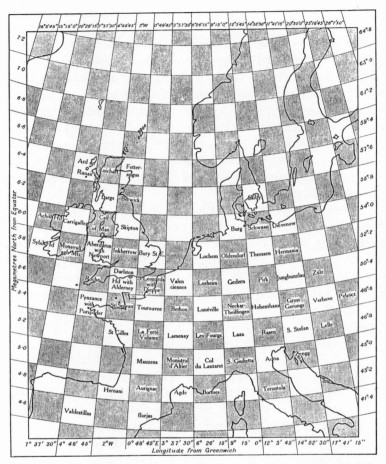

FIGURE 3. The area of forecast covered by Richardson's forecast model. When Lewis Richardson envisioned the development of a forecast model that used numerical equations, he determined that information had to be calculated at each of the squares shown in the figure. His spacing for the squares was at fixed intervals of longitude in the east–west direction, and at fixed distances from the equator in the north–south direction, given on the left border on the map. The corresponding latitudes are shown on the right side of the map. He realized that each box should contain an equal volume and that latitude could not be used as a north–south measure since the latitude lines are closer together as they approach the poles. (From Richardson, 1922, used with permission.)

So to devise a truly accurate numerical model we need data for as many vertical layers as possible. But the use of such data is usually limited by the capabilities of the computers used. Richardson's model had five layers in the vertical, whereas the first model that solved equations of motion that were similar to the basic equations used by physicists (called *primitive equations* by meteorologists) became operational at NMC in the 1960s.[7] The latter model had six levels, and a comparison of the vertical structure of the two models is shown in Figure 4.

Richardson divided his atmosphere into layers of equal mass. In doing so, he chose the divisions between the layers to be at 2.0 (1.2), 4.2 (2.5), 7.2 (4.3), and 11.8 (7.08) kilometers

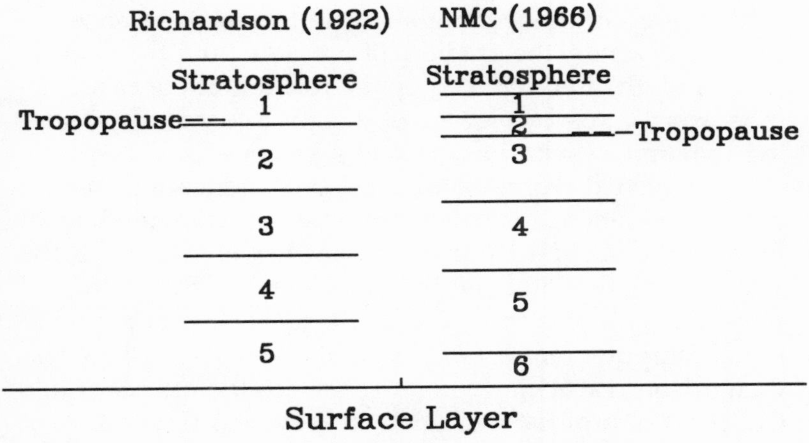

FIGURE 4. Comparison of the vertical structure of Richardson's model in 1922 with that of the operational model used by the U.S. National Meteorological Center (NMC) in 1966. After years of research and development using high-speed computers, NMC's first widely used numerical forecast model had a vertical (and horizontal) resolution that was remarkably similar to what was envisaged by Lewis Richardson more than 40 years earlier.

(miles), the heights at which average pressures of 800 millibars (the units that meteorologists use to measure pressure;
abbreviated mb), 600 mb, 400 mb, and 200 mb are located.
He chose these levels because some "elegant approximations" could be made for the equations that he was using.
What is so insightful in Richardson's work became common
practice in meteorology some 20 years later.

Another key player in the history of numerical weather
forecasting was Carl-Gustaf Rossby, often referred to as the
"father of modern meteorology." The native Swede had been
a professor of meteorology at MIT, an assistant chief of the
U.S. Weather Bureau, and a professor at the University of
Chicago before returning to Sweden in his later years.
Rossby's major contribution, among many, was that he
simplified so many aspects of the forces that play a role in
meteorology into one great struggle to seek a balance between the winds, the gradient of pressure, and the Coriolis
force (i.e., the force created by the rotation of the earth). In
other words, Rossby viewed weather as a constantly changing dynamic system in search of a balance.

Between the time of Richardson's pioneering treatise
and the operational use of the primitive equation model that
became the standard numerical weather prediction in the
1960s, other fundamental breakthroughs in the science of
meteorology took place to ensure the realization of an operational computer forecast model similar to that envisioned by
Richardson. Particularly noteworthy are the efforts of Jule
Charney, a professor of meteorology at MIT and Arnt
Ellassen, a Norwegian meteorologist. Charney system-
atically simplified the intricate equations describing the motion of fluids (called complex partial differential equations)
into expressions that could be programmed and solved by a

computer. Ellassen's work transformed these equations onto a pressure (instead of an altitude) vertical coordinate system, which simplified their calculation. In one sense, this transformation resulted in a vertical coordinate system that was very close to what had been envisioned by Richardson when he described his vertical layers that were composed of equal mass.

The grid that Richardson used for his calculations in 1922 is shown in Figure 5. Each grid box measured 200 km (120 miles) from north to south, and 3° from west to east (a variable distance that depends on latitude but which is also close to 200 km at the latitudes that Richardson did his first forecast). For comparison, the first operational NMC model used a grid spacing of 381 km (228 miles) in its calculations. Most remarkable, however, is Richardson's vision of how massive amounts of numerical computing would have to be done to accurately predict the weather. And this was years before even the most primitive computer.

If you can imagine a beehive, full of busy little bees making honey, you'll have some idea of Richardson's vision. Only instead of bees he imagined people, computing drones, carrying out the mathematics necessary to achieve his numerical forecasts. The equations he used in his models require the solution of several *differential equations* at each point. A differential equation is one that uses a knowledge of differences of various quantities between two points. For example, if we know that it is colder in Toledo than it is in Cleveland at a given time, and we also know that the wind is blowing from west to east, then we can expect that the temperature in Cleveland will drop within a given time span.

But there are more variables in the equation. To accu-

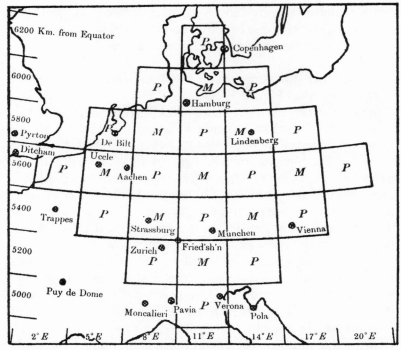

FIGURE 5. The actual grid boxes for which Richardson obtained data and carried out his calculations for his first and only numerical forecast published in 1922. The M's and P's within each box signify the kinds of parameters calculated in each box (M for momentum and P for pressure). He made a forecast at two points: the M point in the center of the region, and the P point directly south of it. The input data were derived from analyses at the surface and at an altitude of ~18,000 feet (a pressure level of 500 millibars) from maps during the period May 18–20, 1910. (From Richardson, 1922, used with permission.)

rately solve the differential equation for Cleveland, we need to know what the conditions are in Cleveland and in Toledo to the west. We need to know the wind conditions at both points and at places in between, like which way the wind is blowing in Sandusky. This is what makes weather forecasting so unpredictable: a forecast at any point on the grid is dependent on what the variables are at *all* points within the computational matrix. Although what is happening in Cleveland may initially depend almost entirely on what is happening in Toledo, eventually, to predict what will happen over a longer span of time, we are going to need to know what the weather is in Chicago, Minneapolis, and points farther west.

The way Richardson envisioned the process, a human computer for each grid cell would be responsible for taking the observations from the field and converting this information into usable quantities for the calculations to follow. Each computer (person) would be furnished logarithm tables of five digits and a modern 10-inch slide rule. Another group of human computers would solve the differential terms of the equations at each grid cell. He estimated that it would take 32 individuals at each point if the computations were done every three hours "just to keep pace with the weather."

Figure 6 shows one of the computing forms that Richardson developed for use in his computing factory. This form illustrates the tedious labor involved in calculating just one variable at one location. This particular form was used to calculate "diffusion produced by eddies," an important component of the transport of air from one grid cell to another. In all there were 22 forms to be filled out at five levels for each variable at each grid cell. In addition, there were forms to calculate:

COMPUTING FORM P VI. FOR Radiation due to atmospheric and terrestrial temperature

The equations are typified by Ch. 4/7/1 # 18, namely $\Sigma\mathcal{E}_s = (1-\bar{\eta})\Sigma\mathcal{E}_g + \eta\bigcirc\theta^4$ for descending radiation
The unit for radiation during the interval of time is throughout 10^5 ergs cm^{-2} = 1 joule cm^{-2}

$\bigcirc = 5.36 \times 10^{-5}$ ergs cm^{-2} sec^{-1}
$\Lambda = 1.48(1+40\mu) \times 10^{-3}$ cm^2 grm^{-1}

Longitude 11° East
Latitude 5400 km North

Time from 1910 May 20d 4h to 10h G.M.T. Interval 6 hours

Leave this until Form P x has been filled in

Height	Fraction transmitted $=1-\bar{\eta}$ $=e^{-\Lambda\sec 55°}.R$ if clear	Radiation emitted to either side $=\bar{\eta}\bigcirc\theta^4$	Downward — Amount transmitted by stratum	Downward — Flux $\Sigma\mathcal{E}$	Downward — Amount gained by stratum	Upward — Amount transmitted by stratum	Upward — Flux $\Sigma\mathcal{E}$	Upward — Amount gained by stratum	Total gain by stratum
h_0	0·581	97	0	0	− 97	301	398	+121	+ 24
h_2	0·583	129	57	97	− 89	390	519	+150	+ 61
h_4	0·586	212	109	186	− 135	457	669	+111	− 24
h_6	0·584	292	187	321	− 158	488	780	+ 56	− 102
h_8	0·588	326	282	479	− 129	510	836	+ 33	− 96
h_L	zero	869*	0	608	− 261	0	869	0	− 261
				869			869		

* At 294·3 abs see Form P x.

Loss to space = sum = − 398

For detached cloud see Computing Form P III
For continuous cloud see Computing Form P I
For R, θ and μ see Computing Form P I

FIGURE 6. One of the computing forms that was used by one of the human computers to perform the calculations necessary for Richardson's forecasts using numerical methods. (From Richardson, 1922, used with permission.)

1. Pressure, temperature, density, water, and cloud input
2. Gas constant, thermal capacities, entropy derivatives
3. Stability, turbulence, heterogeneity, detached clouds
4. Solar radiation at "band" wavelengths
5. Solar radiation at other wavelengths
6. Radiation due to atmospheric and terrestrial radiation
7. Evaporation at the interface
8. Fluxes of heat at the interface
9. Temperature at the radiating surface (part I)
10. Temperature at the radiating surface (part II)
11. Diffusion produced by eddies
12. Summary of gains of entropy and of water
13. Divergence of horizontal momentum, increase of pressure
14. Stratosphere, vertical velocity and temperature change
15. Transport of water
16. Water in the soil
17. Temperature in the soil, latent heat of evaporation
18. Stresses due to eddy viscosity
19. Stratosphere, horizontal velocities and other dynamical terms
20. Dynamical equation for the eastward component
21. Dynamical equation for the northward component
22. Summary of changes deduced from the initial distribution

If there were 2000 active points around the globe at which the calculations had to be done, then it would take

64,000 human computers just to carry out the calculations.
As Richardson wrote

> That is a staggering figure. Perhaps in some years' time it may be
> possible to report a simplification of the process. But in any case,
> the organization indicated is a central forecast-factory for the
> whole globe, or for portions extending to boundaries where the
> weather is steady, with individual computers specializing on the
> separate equations. Let us hope for their sakes that they are
> moved on from time to time to new operations.[8]

Richardson envisioned a physical facility resembling a
theater . . .

> except that the circles and galleries go right round through the
> space usually occupied by the stage. The walls of this theatre are
> painted to form a map of the globe. The ceiling would represent
> the polar regions, England is in the gallery, the tropics in the
> upper circle, Australia on the dress circle, and the antarctic on
> the pit. A myriad computers are at work upon the weather of the
> part of the map where each sits, but each computer attends only
> to one equation or part of an equation.
>
> The work of each region is coordinated by an official of
> higher rank. Numerous little 'night signs' display the instan-
> taneous values so that neighbouring computers can read them.
> Each number is thus displayed in three adjacent zones so as to
> maintain communication to the North and the South on the map.
>
> From the floor of the pit a tall pillar rises to half the height
> of the hall. It carries a large pulpit on its top. In this sits the man
> in charge of the whole theatre; he is surrounded by several
> assistants and messengers. One of his duties is to maintain a
> uniform speed of progress in all parts of the globe. In this
> respect he is like the conductor of an orchestra in which the
> instruments are slide-rules and calculating machines. But in-
> stead of waving a baton, he turns a beam of rosy light upon any
> region that is running ahead of the rest, and a beam of blue light
> upon those who are behindhand.
>
> Four senior clerks in the central pulpit are collecting the

future weather as fast as it is being computed, and dispatching it by pneumatic carrier to a quiet room. There it will be coded and telephoned to the radio transmitting station.[9]

Realizing the need for ever-improving research to be done, Richardson also envisioned an adjacent building housing a research staff whose sole job was to invent such improvements. Any changes devised by the research group would be tested "on a small scale before any change is made in the complex routine of the computing theatre."

In his book published in 1922, Richardson presented one case study for the period May 18–20, 1910, in which he predicted a large storm affecting central Europe. It should be noted that Richardson actually began work on this example within two years of its occurrence, but the start of World War I and Richardson's military duty[10] delayed its completion.

Despite the immense amount of work he put into the study and the subsequent monumental book he wrote to describe his dream of the "forecast factory," the actual forecast was a complete bust! At the central grid point, Richardson's numerical model came up with a pressure change of 145 mb in six hours. Over the spatial scales for which Richardson's model was derived (recall that he used grid 1ells of approximately 200 miles), such pressure changes have never been observed in nature.

Even the most intense hurricane in recent history, Gilbert in 1988, with winds of more than 200 miles per hour, registered a pressure gradient of 120 mb. Furthermore, it took nearly a week for Hurricane Gilbert to reach such a mature state. Richardson's model had an even more intense development take place in the much shorter time of only six hours.

WHY RICHARDSON FAILED

The failure of Richardson's first numerical forecast resulted in the complete halt of further numerical experiments
for more than two decades. Even though the forecast was a
complete bust, no one could explain why it had failed until
much later. In the first place, no one else had ever tried to
solve such a complex set of numerical equations. In fact, one
of the fundamental laws of numerical analysis was not even
published until 1928, six years after Richardson's book was
published. This law states that a fundamental condition for
the solution of the type of equations that Richardson was
trying to solve is called the *computational stability criterion*,
primarily attributed to the German mathematician Richard
Courant. According to Courant, the intervals of time must
be short enough that transport from one grid cell to another
can be resolved. That is, since weather is so dynamic, the
variables can change momentarily and this must be taken
into account when resolving the equations.

In Richardson's case, his use of a grid cell of 200 miles
and a time step of six hours would violate this condition if
any winds on his computational matrix exceeded 33 miles
per hour. The reason for this is that the differential equation
at a particular grid cell depends on the information available
on either side of the cell and assumes that the measurements
at those cells, as well as at the cell for which the calculation
is being carried out, contain information for the same time.
If the wind is too fast, then the information for the air
supposedly at the adjacent cell for a particular time hasn't
reached that adjacent grid cell yet. Therefore, any parcel of
air traveling more than 33 miles per hour would not have
reached the adjacent cell at which the measurements are
needed to solve the differential equation. In the upper levels

of the atmosphere, near the jet stream, wind speeds of more than 100 mph are common, so, in retrospect, it is not surprising that Richardson's calculations proved false.

Another primary reason for Richardson's failure was that he attempted to solve for too many parameters in his calculations for the scale of motion he was trying to forecast. This shortcoming is one of the fundamental and most challenging problems in the science of meteorology. Resolution of scale has always been a thorn in the side of atmospheric scientists because there is only *one* atmosphere. Yet within that one atmosphere are forces that contribute not only to the formation of an isolated thunderstorm but also to the formation of the intense cyclone in which many thunderstorms are embedded. The difficulty arises when we try to differentiate what is happening by our measurements at a particular location.

For example, an observation made at point A, the site of a severe local thunderstorm, may include reported surface winds from the north of 50 mph. The circulation around the thunderstorm, however, may only extend 10 miles or so. At the adjacent point B in our computational matrix, 200 miles away, the situation may be very calm and representative of the conditions over the entire grid cell. The success of a numerical calculation assumes that the observations made at a particular location are *representative* of the conditions over the entire grid cell. In other words, the presence of the thunderstorm at point A at the time of its observation would produce extremely inaccurate data for use in the calculations for the larger scale of the grid cell.

It was Rossby in the late 1930s who simplified the complex partial differential equations Richardson was attempting to solve by showing that many of the terms in these equations could be neglected, since they were not relevant to the large scale on which such phenomena as cyclones were dominant.

Shortly after the unveiling of ENIAC following the end of World War II, John von Neumann organized the Electronic Computer Project at the Institute for Advanced Study (IAS) in Princeton, New Jersey. The goal of the project was to design and build an electronic computer that would greatly exceed the power and capabilities of the existing electronic computers, such as ENIAC. In 1948 Jule Charney established the Meteorology Group within the ECP. Their task was to apply dynamical laws, using the electronic computer to be developed by the Project, to the problem of forecasting the weather.

There were two features of the new computer that would change the way future computers were to be used: one was a function that would allow computer programs to be stored within the computer (what we now commonly refer to as "software"). The other feature was one that would allow parallel processing, the ability of the computer to compute two tasks at the same time. This latter feature greatly enhanced the amount of work that could be done by the electronic computer. For example, Charney successfully ran a computational experiment on the ENIAC in which he produced a 24-hour forecast in 24 hours of computational time. Obviously, by the time the computer was through with the computations, the forecast was obsolete. On the new IAS computer, however, the same numerical experiment used only five minutes of computer time. The numerical model used in this experiment consisted of only one vertical level at 500 mb, which lies at approximately the middle of the troposphere, about 5200 meters (18,000 feet). The model was capable only of forecasting pressure changes at the center of that level.

To achieve greater accuracy and higher predictability, researchers then turned to multilevel models. One such model, which included three levels, was developed in 1950 by IAS researchers. For their observations they focused on the Thanksgiving holiday period, November 23–26. Their calculations predicted a storm. Sure enough, a large storm moved into the northeast just in time for the holiday weekend. The storm was greeted with enthusiasm, for it verified that at last a computer could calculate an accurate forecast, and it took only 48 minutes. Richardson's dream of nearly 40 years earlier was getting closer to becoming a reality; only this time one machine was doing the work of an entire army of human calculators.

The success of the IAS model provided the impetus for the formation of the Joint Numerical Weather Prediction Unit (JNWPU) in 1954, which consisted of representatives from the civilian National Weather Service, the Air Force's Air Weather Service, and the Naval Weather Service. By 1955, using an IBM 701 computer, the JNWPU was issuing numerical weather predictions twice daily.

Not everything was perfect with the new computers and the new predictions. First of all, the early numerical forecasts still were not as good as those done by the human forecasters on duty. One of the decisions that had to be made by the JNWPU was what kind of numerical model was most usable for forecasting the weather. In the intervening decades since Richardson's work, the field of meteorology had made remarkable advances, both in theory and in practice.

One advance was a direct result of World War II: the growth of a wide-ranging observational network. This network was composed of observing stations located at airports, air bases, and weather bureau stations. Such complete observing stations had not been available back when Richardson had first conducted his experiment. But by the 1950s, with the network in place, there was no shortage of data to feed into the equations to be computed.

The JWNPU had a choice: it could keep on studying the mechanics of the atmosphere and hope to eventually demonstrate the utility of numerical weather prediction as an operational product or it could proceed immediately and help the Weather Service generate forecasts using numerical weather prediction in the field. The difference meant valuable time before a system would become operational. The committee finally chose the latter, but then had to decide how to develop such a model. Should they set up an operational numerical model that tried to solve all the equations that Richardson tried to solve (these models are now referred to as primitive equation models)? Or should they utilize the theoretical knowledge of Rossby and his research to produce a simpler model that could be run more efficiently on an electronic computing machine?

Fortunately for us, they chose to use one of the simpler models, one that had been developed by Jule Charney at Princeton. Charney's model had already been tested on about a dozen case studies. After the model was programmed for the IBM 701 in 1955, it was run on an operational schedule, and subjected to the critical eye of the practicing meteorologist in real time. Not only were the results unable to predict the weather reliably and accurately, but the output of the calculations provided no useful information to the forecaster, either.

Although disappointing at the time, this "baptism by fire" approach was the key to the eventual success of operational numerical weather prediction since these quick results immediately brought together the talents of two groups of meteorologists who until then had worked separately: the modelers who developed the models and the practicing meteorologists in the field. Instead of both groups of scientists going their separate ways to analyze and hope for a solution, they worked together to bring about a product that eventually proved beneficial. Without this cooperative effort the meteorologist in the field would have felt threatened by the numbers being generated by the newfangled computing machine. The payoff came in 1958 when the problems were identified and solved, a suitable automatic analysis system was invented, and automatic data handling was developed. As a result, timely numerical forecasts were delivered to forecasters, who in turn used them as *guidance* for their own manually produced prognostic charts. In other words, the computer became another tool for the "art" of weather forecasting. Even to this day, many of the meteorologists in the field still believe that weather forecasting is part science and part art.

The graph shown in Figure 7 illustrates how numerical weather prediction has improved from 1955 through 1988. The quantities plotted each year are a statistical measure of success for a 36-hour forecast called skill scores. Without going into a technical discussion describing exactly what these numbers represent, let us say that after many years of trying to come up with a way to measure the success of a forecast, these numbers define how much the forecast has missed predicting the pressure of the atmosphere at a given height over the entire area of computation of the model which includes North America and the adjacent ocean areas. On this particular scale, a score of 20 would be a perfect forecast, whereas a score of 70 would be defined as worthless.

The impact of computer-generated forecasting models

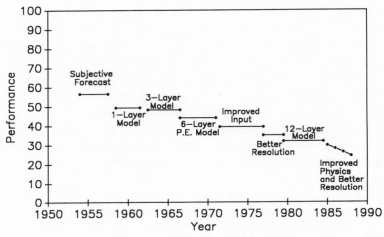

FIGURE 7. Performance of the NMC numerical models.[12] A measure of the performance capability of numerical models covering more than 30 years of research and implementation is shown in this figure. As computers get bigger and faster, these models can employ higher resolution (more vertical levels and smaller distances between points of calculations) and more detailed and complicated equations describing the physical processes that drive the weather. The result has been a gradual improvement in the forecasts.

can be seen in this figure by the drop in the skill score beginning in 1958. About two years following this success, the JNWPU was divided into three organizations which still exist today: the National Meteorological Center (National Weather Service), the Global Weather Central (the U.S. Air Force) and the Fleet Numerical Oceanography Center (the U.S. Navy).

By 1960, numerical weather prediction had progressed to the extent that some of the forecasts not only surpassed in quality those made manually at the National Meteorological Center (NMC), but also could not successfully be improved by manual methods. Automated *products* (the term given to the forecasts by the weather services) began replacing manual products about that time. But the road to success was not without its obstacles.

As a matter of fact, the scientists who had devoted so much effort to the development of numerical prediction models had to take a step backward to obtain the acceptance of the meteorologists out in the trenches who had to provide the public with a weather forecast. Although a three-level model developed by Charney was believed to be the best available for forecasting at the time, the operational meteorologists were actually intimidated by it because of its sophistication. As a result a relatively simple one-layer model was used to provide guidance. Even though the researchers on the leading edge of model development knew that this one-level model did not provide as good a forecast as the three-level model, it was important for the practicing forecaster to understand what these models were forecasting, and why more complicated models were necessary to provide good results.

Only when the numerical solutions of the equations that describe the relatively simple physics of this model were understood could the dream of numerical weather prediction proceed at a steady pace, even if that pace was slower than originally envisioned. For example, these simple models could solve equations that described how winds had to adjust to the migration of a pool of cold air and its accompanying high-pressure system into the forecast area. But the more complicated physics that took into account the resultant change of the vertical distribution of temperature and how these affected the weather could not be described in these first one-layer numerical models. Not too long thereafter, however, the three-layer model was implemented as the operational tool of the forecaster; eventually, by the late 1960s, an even more complicated numerical model, known as the primitive equation model using six layers, a model similar to the type that Richardson had envisioned, became the operational tool at the National Weather Service. Whereas all the numerical models prior to that time had only produced reliable forecast statistics for the middle of the atmosphere (which is where the performance score in Figure 7 had been determined), the six-layer model was the first one to produce a prediction at sea level that was directly competitive with those of the synoptician.

The evolution of computers and meteorology seemed to have come full circle with the development of the six-layer model so similar to the model of Lewis Richardson 50 years earlier. Using this model, a substantive effort was made during the 1970s to improve the method for entering data into the model—a process called *initialization*. The initialization process consists of entering the observed data into the computers, but including certain physical constraints on the calculations. For example, a particular weather observation station may be situated in a valley that is susceptible to locally heavy fog and therefore may not provide surface temperatures that are characteristic of the larger region. Eventually this fact, obvious to any trained field meteorologist, had to be programmed into the computer so that the impact of the very localized temperature difference observed at this station would not create havoc when entered into the computer calculations.

Another important aspect of the initialization process is to make sure that all the data fields are consistent with each other. For example, using the observations at a particular time, the distribution of temperature can be determined. Using the same set of observations, a pressure distribution or an analysis of the winds can also be readily derived. Even though the temperature field, the winds, and the pressure distribution may all depict the reality of the situation, they may be slightly inconsistent with each other insofar as the physics within the model is concerned.

Here's one example of how computer forecasts can become erroneous if the input data are not initialized. Suppose the air pressure at point A in the computational matrix is off by one millibar and the pressure at point B is also off by one millibar in the opposite direction. One of the real basic laws of physics is that motion takes place when an imbalance of pressure is present. We can observe such a phenomenon by watching a block of wood float on top of a tub of water. Suppose that the block is perfectly still in the water. The pressure exerted on the bottom of the block by the water is completely balanced by the pressure on the top of the wood exerted by the atmosphere. Similarly, the pressure on all sides of the block are equal so the block does not move. But let's see what happens if you start blowing on one side of the block—another way of saying the pressure on one side of the block has increased. (The same effect would be accomplished if you added that pressure by pushing on one side of the block with your finger.) The result is that the block moves with a certain speed. The speed of that movement could be calculated if you could quantify the force exerted on the one side (note that if the same amount of force were applied on the opposite side, there would be no motion). So the motion of the block can be obtained accurately if the pressure distribution on it is known accurately.

Now, let's go back to our original problem of the pressure being only slightly inaccurate at both points A and B. Since one of the driving forces behind the forecasting of wind speeds is the constant struggle to find a balance between the pressure gradients within the computational matrix, the calculated winds by the numerical model will automatically predict incorrect winds if the pressures are not quite correct. Even these slight inaccuracies eventually spread throughout the entire forecast region and the resultant forecasts will be totally incorrect.

So although such initialized errors seem small, they are capable of greatly influencing the accuracy of a forecast. To get around such problems, scientists have spent a considerable amount of time developing initialization schemes. Essentially, a forecast is made for a short time forward and the results are then run backward to the time at which the observations had been taken. These backward forecasted winds, temperatures, and pressures are then used as the input for the operational models to forecast the future. Even though the initialized input data were very similar to what was observed, the slight differences greatly improved the forecasts.

By the late 1970s computer power had increased enough to enable forecasters to cut in half the computational grid, reducing the grid cell to 191 km. This new six-layer primitive equation model was uncannily close to the five-layer, 200-km grid model that Richardson described in such detail in 1922.

THE AGE OF HIGH-SPEED COMPUTING

As computers became even more powerful in the 1980s, weather forecasts were able to improve because more complicated physics and higher resolution could be easily incorporated into the models. That is, the improvements made in the past decade or two are primarily a result of faster computers able to handle more complex terms in the equations that describe the physics behind the formation of weather.

Since 1952 computer technology breakthroughs have been so common they've become the norm. The National Meteorological Center, for example, has undergone six state-of-the-art supercomputer acquisitions in its first 33 years of operation—from the IBM 701 to the Cyber 205 in the late 1980s; in between, the NMC used the IBM 704, IBM 7094, CDC 6600, and IBM 360/195. Each new supercomputer has been about six times faster and has about six times the storage of its predecessor, so that the Cyber 205 has about 10,000 times the capacity of the IBM 704 in both its speed and its storage capacity.

Despite these technological advances, the accuracy of the forecasts produced from numerical weather prediction has not increased by orders of that magnitude. By most measures, forecaster skill has more than doubled since the 1950s. This difference between the amount of computing power and the improvement in our ability to produce a more accurate forecast arises from the nature of the problem and its overwhelming complexity. Increasing resolution by doubling grid resolution requires an order of magnitude increase in computer storage capability since grid resolution must be accounted for in three dimensions.

In a 1989 article in *Weather and Forecasting*, Dr. Frederick G. Shuman, a former director of the National Weather Service, wrote that "doubling the resolution of a forecast only improves the skill of the forecast by about 15%, and even then only if the other characteristics of the model are suitably enhanced."

Each new acquisition of a more powerful computer enabled forecasters to use more sophisticated numerical weather prediction models. When the National Meteorological Center obtained the CDC 6600 computer in 1966, it was finally able to inaugurate the use of a primitive equation model of the type that had first been envisioned by Richardson more than 50 years earlier, even though the spatial resolution was not as good as Richardson had envisioned. When the new IBM 360/195 came on line in the mid-1970s, the model resolution was improved to what Richardson had used to make his first computations.

The computer gains made during the last two decades have enabled weathercasters to provide the colorful analyses seen on the Weather Channel, and local television weather forecasts. The powerful computers (now commonly called workstations) that now easily fit on a desktop allow for the animation of geostationary satellite images and composite radar depictions from all stations around the United States. Such color graphics could not have been generated using even the largest computer available in the 1970s.

Until recently, the bottleneck to the development of more accurate forecasts was the lack of a computer powerful enough to handle the data involved in numerical weather prediction. Now we have the powerful computers, but we need to use higher-resolution data. To gather such data using conventional weather instruments is far too expensive. Fortunately, there is a solution. While computer power was growing by quantum leaps, scientists were busy developing a new generation of remote sensing instruments. These instruments are now ready to be launched into space aboard a new generation of satellites. A sky full of satellites aiming sensing devices back through the atmosphere will provide ample work for the high-powered computers on the ground, which should improve numerical weather prediction to a point unimaginable back when Richardson first formulated his model.

FOUR

How Satellites Changed
the Face of the Earth

April 1, 1960 dawned clear and warm over Cape Canaveral, Florida. At 6:40 A.M. the feeding herons in the nearby marsh were startled by a thunderous roar as a Thor-Able rocket leapt from its pad at the U.S. Space Center. Arcing northeastward over the Atlantic Ocean, the rocket pushed its 270-pound, hatbox-shaped spacecraft into orbit 450 miles above the earth. In the nose cone of the rocket was the United States' first tentative step at changing how weather forecasts were made.

The payload on the Thor-Able rocket was nothing glamorous. It was not a spaceship destined for other planets. It was not some fancy device that would change the balance of power in the cold war. But it was a first step in a series of developments that would eventually change the science of weather forecasting forever. What was in the nose cone of the Thor-Able rocket was the first Television and Infrared Observation Satellite, known as TIROS-1.

The launching of TIROS-1 was the culmination of a series of events dating back to the end of World War II. The goal was to extend the eyes and ears of weather forecasters all the way into space, where it could look back at earth with an omniscience barely dreamed of. But the process started way before we had the ability to fly. The process started here in the United States, when the government realized that forecasting weather was not only necessary but essential to an expanding nation.

By the time of the Civil War, the earliest beginnings of a network of weather observing stations was in place. These were the U.S. Army hospital surgeons who, under a government directive, were required to keep climatological and meteorological records at their hospitals across the country. The Army surgeons were joined by telegraph operators as the telegraph spread across the expanding country. This made for a relatively vast network of weather observing stations, each feeding data on temperature, humidity, and winds to Professor Joseph Henry, then president of the Smithsonian Institute. Professor Henry used the data from more than 500 stations to formulate the country's first weather map.

By today's standards, such a map was primitive, relying on surface observations. Even back then, forecasters realized the necessity of upper-air observations. The Civil War disrupted the network, and attempts to reinstate it failed following the end of hostilities. By 1870 a series of private weather observing networks was in place in various areas of the country, but it was obvious even then that the expanding country needed a nationwide network of observing stations if it was to be able to accurately forecast weather.

Since the U.S. Army already had in place a network of telegraph stations, the job of designing a nationwide weather service was given to the Army's Signal Services, later called simply the Signal Corps. Under the umbrella of the War

Department, the Army furnished its telegraph stations with the equipment available at the time: a wind vane, anemometer, rain gauge, thermometer, barometer, and sling psychrometer to measure humidity.

Still, the weather observations were limited to surface conditions. Any research on clouds, for example, was carried out from mountain stations. The Wright brothers had yet to launch the age of flight, which would allow weather instruments to be carried above the ground and into the atmosphere.

By the turn of the century the forerunner to the National Weather Service was moved from the Army's jurisdiction to the Department of Agriculture and named the U.S. Weather Bureau. In 1895 Professor Charles F. Marvin, who would later become the weather bureau's chief, saw the need for some means of obtaining information from the atmosphere above the surface. So he began experimenting with an ancient plaything known to children around the world—the kite.

Marvin devised an instrument to be carried aloft by the kite, which became known as the Marvin kite meteorograph. It could measure temperature, relative humidity, and wind speed. The use of kites to measure the atmosphere up to 1,000 feet became standard procedure at weather bureau stations across the United States. Even after balloon and aircraft measuring devices were introduced, kites remained in service until 1933.

Unfortunately, kites had their problems. To get a kite to an altitude of 1,000 feet required wind speeds of 10 to 15 mph. On calm days the kite couldn't be used. Another problem was rain. A wet kite, no matter how strong the wind, proved too heavy to stay airborne.

Meanwhile, other researchers, aware of the shortcomings of the kite, had been experimenting with balloons.

Balloons had been used before kites. Small, tethered balloons were let up and then observed from the ground to determine upper-level wind speed and direction. These balloons could sometimes reach as high as 6,000 feet. Untethered balloons were also launched and then tracked optically to determine wind speed and direction. Some of the untethered balloons reached altitudes of 10 miles.

By the beginning of this century weather instruments had become more and more refined, so scientists began loading these instruments onto balloons. But even these sounding balloons had their problems. Each balloon carried a meteorograph and a parachute. The balloon was designed to burst at a given altitude, launching the meteorograph back to earth via its parachute, where scientists could analyze the findings. The problem was that sometimes it took weeks for the meteorograph to be found and its contents analyzed. Obviously, by the time the readings were usable they were past history. Sounding balloons were not made a routine part of weather bureau procedure.

By the time of World War I the Weather Bureau tried to expand its upper-air observations by increasing the use of kites and balloons. It attempted to persuade Congress to release more funds so it could expand to the use of airplane observation. The Congress of 1920 opposed such a plan.

Five years later, however, the Weather Bureau joined up with the Navy to conduct observations from airplanes flying from the Naval Air Station at Anacostia, in Washington, DC. Every morning at 8 A.M. a plane would take off with a Marvin meteorograph atop the upper wing. It would fly to 10,000 feet and return. The meteorograph was sent to the Weather Bureau along with the pilot's personal observations.

This technique proved so successful that by 1931 airplane observations were routine, with planes taking off under Weather Bureau contract from airports in Cleveland, Chicago, Omaha, and Dallas. The pilots of such planes were paid $25 by the Weather Bureau for each flight. The plane had to reach a minimum altitude of 13,500 feet. For each 1,000 feet above that the pilots received a 10% bonus. By 1937, airplane observation flights were launched regularly from 30 fields across the United States, operated by Army, Navy, and civilian pilots under contract to the Weather Bureau.

By 1938 the minimum altitude required of these pilots was 16,500 feet. Lack of oxygen and other flight hazards accounted for 12 deaths in the program between 1931 and 1938. Advances in airplane technology and the imminent danger of war soon forced the military to pull out of the program. The meteorograph had been easily mounted atop the biplane, but with the advent of the monoplane as the plane of choice in the military there was not as much room to accommodate the weather instrument. By the time World War II began, such flights had been discontinued.

DEVELOPMENT OF THE RADIOSONDE

By the time of World War II, the newest electronic advances in radio made such airplane flights unnecessary. Scientists no longer had to search for, locate, and recover the meteorograph after every flight. Now the meteorograph could be launched with a balloon and, by using radio transmission, the data could be sent instantaneously back to the ground station. No longer would scientists have to track the balloon by sight and then retrieve the instrument. Data would be sent as the balloon continued its flight.

The development of what later became the radiosonde (a radio sounding balloon device) was not an overnight occurrence. As far back as 1928, the Russian scientist P. A. Moltchanoff had perfected a radio meteorograph. In 1931 he had launched three such balloons from the dirigible Graf Zeppelin while it cruised above the Arctic Ocean. The balloons had worked, sending data back as they soared above the cold seas. The success of such radio meteorographs spurred more research in other countries, including the United States. By the time of World War II, the armed forces were making extensive use of such balloons to meet the demands of combat.

Unfortunately the radiosonde had its shortcomings. The instrument aboard the balloon could measure only temperature, pressure, and humidity. To measure the wind, other techniques had to be implemented which involved visual tracking of the balloon so observers could calculate the wind speed from the balloon's azimuth and elevation. This worked fine except when the balloon went above the clouds or there was fog on the surface. What was needed was a truly all-weather radiosonde that could be tracked no matter what the conditions.

Scientists developed a two-pronged solution to the problem. They implemented a direction finder on the radio signal coming from the balloon and they utilized the newly developed miracle of radar to track the balloon from the ground. A radiosonde-equipped balloon from which wind information is obtained by either of these means is called a rawinsonde.

In retrospect, it is easy to see how the country's efforts during World War II produced technology that could benefit people during peace. The development of radar, for example, was a major impetus in the growth of the postwar civilian airline industry. But perhaps the most exciting development to come out of the war, and the one with the greatest impact upon weather forecasting, was rocketry.

THE DEVELOPMENT OF ROCKETRY

Back in the 1920s, encouraged by Dr. Robert H. Goddard, scientists had begun to conceive of using rockets to sound the atmosphere. But the idea had been ahead of its time. It took Nazi Germany's development of rocket technology, exemplified by the V-2, before concept and execution caught up with each other. The V-2 rocket was 46 feet long and weighed 14 tons. It could be launched at ground targets as far as 200 miles away. The apex of its arc reached 55 miles above the earth's surface. The V-2 was a giant leap forward in rocket technology.

By the time World War II ended, there were plenty of V-2 rockets, parts, and experts roaming around the postwar scene. This provided the United States with the opportunity to use the new technology for vertical soundings of the upper atmosphere. The first German V-2 rockets were launched in 1946 by the U.S. Army at the White Sands Proving Ground in New Mexico. Over the next six years more than 50 V-2 rockets were launched, reaching altitudes of more than 100 miles. By the early 1950s U.S. scientists had improved the original V-2 and developed the Viking and Aerobee rockets to replace it.

Shortly thereafter, the U.S. Navy began using rockets for airborne research, launching them with motion picture cameras on board to film the earth from high altitudes. This new development gave meteorologists, for the first time, a picture of the earth from beyond the atmosphere. Seeing earth from space proved fascinating, but also frustrating. What good were motion pictures of the earth's surface taken hours or even days before? Scientists had faced the same problem with balloons in the early days: they had to wait until the rocket fell back to earth to retrieve the films. That was time-consuming and expensive. What was needed was a vehicle that would take pictures of the weather from rocket altitudes but that wouldn't immediately fall back to earth.

THE BIRTH OF SATELLITES

In 1954 the idea of satellite-based observation of the earth's atmosphere got a boost when a Naval Research Laboratory Aerobee rocket climbed about 100 miles above White Sands to bring back the most dramatic and meteorologically significant pictures yet obtained by a sounding rocket. The two reels of 16 mm film revealed a well-developed tropical storm centered over Del Rio, Texas. The storm had not been detected by the conventional weather observing network in place at the time, yet there it was, plainly visible from an altitude of 100 miles.

The discovery of the tropical storm over Texas convinced the U.S. government that some kind of routine meteorological observation from higher altitudes was necessary. In 1955 the first official step was taken to put a satellite in orbit around the earth when the government established the Vanguard program as its contribution to the International Geophysical Year. For the next three years scientists worked diligently to prepare the way for a series of Vanguard launches. The U.S. effort was spurred on by the launching of the world's first satellite, the Sputnik, by the Soviet Union in 1957.

But by the mid-1950s there was another technological advance just waiting for a chance to go up in space. Television had become more than radio's little brother. By the middle of the decade the country was blanketed with television signals. So it was a logical step to start thinking about using television cameras to simultaneously broadcast back pictures of the earth and its atmosphere.

In 1956 the Rand Corporation, under contract to the Army Ballistic Missile Agency at Redstone Arsenal, began work on just such a project, dubbed the JANUS. They came up with a slender, rod-shaped spacecraft that weighed a mere 20 pounds, measured about 42 inches in diameter and about $2\frac{1}{2}$ feet long, and came equipped with a single television camera.

Over the years, as advances in rocketry kept pushing the designers to change their design, spacecraft grew larger and more complex.[13] By the time the project had been taken over by the newly created National Aeronautics and Space Administration at the Goddard Space Flight Center, the satellite weighed over 260 pounds and was $3\frac{1}{2}$ feet in diameter. On board were two compact vidicon television cameras, one with a wide-angle lens and one with a narrow-angle lens. Two separate recording systems were attached to each camera in addition to a control system that could beam the pictures back to earth. To ensure that the two cameras would always be pointing back to earth, the satellite was equipped with two small solid rockets that could be manipulated by ground commands to keep it pointing in the proper direction. The entire electronic setup was powered by batteries that could be recharged by solar arrays that converted the sun's energy to usable electrical power.

It was this latest version, named TIROS-1, that took off on April 1, 1960. It was the world's first meteorological satellite, and it proved once and for all that it was now possible to take pictures of the earth's cloud cover over much of the planet. The goal of TIROS-1 was to demonstrate the feasibility of the entire concept of using satellites to gather information on the earth's atmosphere. By using a slow-scan miniature camera in an earth-orbiting platform that was always pointing in the right direction (toward earth), scientists were able to view the earth from a vantage point that until then had only been imagined. The satellite orbited at an altitude of 450 miles and covered all the land and ocean areas between 48° north and south latitudes as it circled the earth every 99 minutes. The pictures sent back from TIROS-1 on its first orbit changed the field of meteorology forever.

But there were problems. TIROS-1 was spin stabilized, like a gyroscope. It spun at 8 to 12 rpm, which meant the cameras on it variously looked directly or glancingly at the earth's surface. By the time of the launching of the second TIROS satellite, in November 1960, scientists had figured out how to improve the satellite's "attitude" to earth, i.e., the angle at which the satellite faced the earth. The TIROS-2 satellite was equipped with a magnetic coil that provided some control over its orientation.

TIROS-1 proved that a satellite could be placed into orbit and beam back pictures of the earth's atmosphere. But sending back pictures of the earth was just the beginning. Now that they had proved the feasibility of using satellites to collect information in the visible spectrum of the electromagnetic field, why not go beyond the visible spectrum? TIROS-2 tried to answer that question.

On board the TIROS-2 were instruments that could detect infrared radiation, in addition to the two television cameras. Thus the new satellite could measure the emitted and reflected thermal energy of the earth's surface and atmosphere. By being able to observe the earth's atmosphere and surface at different wavelengths within the infrared range, scientists were able to learn much more about the atmosphere than simple visual observation would allow.

For example, infrared sensors set to pick up radiation at a wavelength of 8 to 12 microns (a micron is one millionth of a meter) can detect the so-called "water vapor window." They can "see" all the way to the earth's surface from space unless clouds are in the way, in which case they "see" the tops of the clouds. The energy emitted from the tops of these clouds is an indication of their temperature, which in turn is an indication of the height of the clouds. So by viewing the earth through infrared sensors, scientists could determine not just where cloud cover occurred, but what kinds of clouds and what kind of associated weather went with them.

There were eight more TIROS satellites launched during the first half of the 1960s, all of them successful. Still, the TIROS program was experimental. Its purpose was to prove that it was possible to observe the earth's atmosphere and meteorology from an orbiting satellite. With ten successful launches in five years, the addition of infrared sensors, television tape recorders, and other advanced technological instruments, and an impressive collection of data that changed the way weather could be forecasted, scientists and government officials decided it was time for the next step: initiation of a complete operational system.

SATELLITES BECOME OPERATIONAL

During the TIROS program, most of the collected data were used for research and not as part of the National

Weather Service's daily operations, although weather service operations benefited from the satellite observations. But with the research phase over, what was needed was an operational system for globally observing the earth's cloud cover on a daily, routine basis.

In February 1966 the world's first operational weather satellite system was placed into service with the launching of the Environmental Science Services Administration satellites, known as ESSA-1 and 2. The two satellites served different functions. ESSA-1 provided simple data readouts from Advanced Vidicom Camera Systems (AVCS) while ESSA-2 was equipped with redundant Automatic Picture Transmission (APT) cameras that transmitted actual pictures back to earth. Now scientists had a steady supply of data and pictures coming from space as the satellites orbited.

The TIROS Operational System (TOS) worked by receiving the pictures from the satellites at its Command and Data Acquisition (CDA) stations in Virginia and Alaska, then relaying the data to the National Environmental Satellite Service at Suitland, Maryland, where they were processed and disseminated to forecasting centers around the world.

By the end of the 1960s the TIROS Operational System was operating smoothly, sending back data and pictures to the National Weather Service. In 1970 the NWS became a part of the newly formed National Oceanic and Atmospheric Administration (NOAA), which replaced the Environmental Science Services Administration. NOAA's first launch was ITOS-1 (Improved TIROS Operational System), on January 23, 1970. In addition to being able to send back data and pictures, the new satellite carried an operational two-channel Scanning Radiometer, which provided day and night radiometric data that were sent back to earth. With the newer generation of satellites, scientists could observe the global atmosphere every 12 hours from a single orbiting satellite.

The use of such satellites was a great improvement over ground-based weather observations. For the first time meteorologists could see the weather patterns over their areas. The view of the earth from space was quite remarkable. The instruments aboard the spacecraft were sending back valuable data that improved weather forecasting substantially.

But there was still room for improvement. One of the problems meteorologists were facing was the satellite's orbit, taking it around the earth in about 12 hours. This allowed for a broad picture of the earth's atmosphere, but did not allow for a penetrating look at more local conditions. So in 1966 NOAA initiated work on the first Geostationary Operational Environmental Satellite (GOES), to demonstrate communications technology by use of a satellite in a geostationary orbit.

In this case a satellite is launched into an orbit so that its speed is about the same as the rotating earth. This enables the satellite to remain over a given geographical point while orbiting in a west-to-east direction at an altitude of a little over 22,000 miles. The development of the first three satellites, called the ATS series (Applications Technology Satellite), led to the development of the first satellites designed specifically for meteorological purposes. These include the later GOES series of satellites that continue to orbit today and are responsible for most of the satellite views we see on nightly television.

But the GOES satellites were limited to views of the earth between roughly 60° N and S. To complete the view of earth, a series of polar-orbiting satellites is also operational.

Another important advantage of the polar-orbiting satellites is that they orbit the earth at much lower orbits, typically 400 to 500 miles. Because of their relative closeness, they can see much better and provide higher-resolution data. The current NOAA polar-orbiting (called NOAA-11, NOAA-12, etc.) satellites are descendants of the ESSA-1 satellite launched into polar orbit in 1966. They provide data with a 1-km resolution at five different spectral portions of the electromagnetic spectrum. We will return to how the information from these satellites is currently used for forecasting and for research purposes later in this chapter.

It is the GOES satellite (see Figure 8) that most people are familiar with, since it is this satellite's pictures that provide the basis for most weather forecasts in the United States. The satellite itself is a vertical cylinder about 20 feet long, covered with solar cells and containing both visible and infrared detectors that record and transmit data back to earth on a regular basis.

This space view of earth allows forecasters on the ground to "see" weather systems and to track them, not only across the continent, but also across the entire globe. Global coverage is now a reality as the Europeans and the Japanese also have geostationary satellites that are used by countries in those regions of the world for their weather forecasts. With orbiting and geostationary satellites keeping track of our weather from space, and more than 120 countries using the data supplied by such satellites, the dream of scientists back in 1960 to establish "an operational system for viewing the atmosphere regularly and reliably on a global basis" has been realized. Satellites can not only detect large storm systems across the globe but also can now determine moisture content of air masses, monitor worldwide vegetation levels, assess damage to the environment, determine ocean

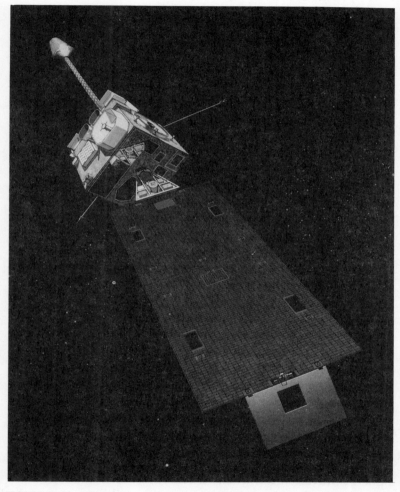

FIGURE 8. One of the Geostationary Operational Environmental Satellites (GOES) that will compose the GOES-NEXT series of satellites. (Courtesy of NASA.)

surface temperatures, and determine extent of snow cover. By 1974, the field of satellite meteorology improved to such an extent that NOAA was able to produce satellite loops. By piecing together a geostationary image every 30 minutes, we are able to view the development of weather systems as they happen. By the 1980s, we've come to expect to see these loops as part of a regular newscast.

Such uses have found favor with private and public agencies and any information gathered by satellite is now available to any individual owning a computer. This has enabled the development of a new kind of "weatherperson," the private weather forecaster. A private weather forecaster is able to tailor his or her forecast to a specific localized area or for a specific interested party, such as ski resorts, small towns, and building contractors.

THE FUTURE SATELLITES

To get a feeling for the future of weather satellites, we need to know how these satellites "work" and what is planned to make them work better. Satellites gather information through the use of several sensors. Sensors are devices that gather information about the atmosphere at specific portions of the electromagnetic spectrum. These devices are able to detect radiation at these particular portions of the spectrum and then derive quantities such as temperature and humidity. These sensors are mostly sensitive to the infrared portion of the spectrum, and they look at specific wavelengths in the infrared. This is analogous to looking at only very specific color in the visible portion of the electromagnetic spectrum since each color is determined unambiguously by the wavelength at which it emits its energy. Thus, the sensor would only be sensitive to a particular color of light and by quantifying how much of this radiation (i.e., color of light) reaches the sensor, the amount of water or the temperature can be derived from a specific location. By aiming these sensors at the earth from space, scientists can get a "picture" of such things as moisture levels at particular altitudes of the atmosphere, cloud depth and content, and so forth. Obtaining a picture (or image) is achieved by scanning across the field of view of the terrestrial scene. West-to-east scan lines are formed by the rotation of the spinning spacecraft. The lines are moved by a stepped mirror to cover the north-to-south aspect of the image. It takes about 20 minutes to scan a complete image. When the system is operating at full capability, there are two satellites stationed above the earth at the same time. Although both satellites sit on the equator, GOES-EAST is located at 75° W longitude and GOES-WEST is located at 135° W (see Figure 9). During most of the year, the weather systems that impact the United States move from west to east and the GOES-WEST satellite is the one that is used for watching these storms develop before they reach the mainland. During hurricane season, which is officially defined as June through November, the GOES-EAST satellite is more valuable for forecasting approaching storms that are often seen in their embryonic state as they move off the coast of northern Africa. During times when only one GOES satellite is operational (because one has failed for one reason or another), the Weather Service will move the remaining satellite using small rocket engines on the satellite, to the east during the hurricane season, and to a more western location for the rest of the year.

Two full earth disk images are provided by each satellite each hour. During daylight hours, these images alternate

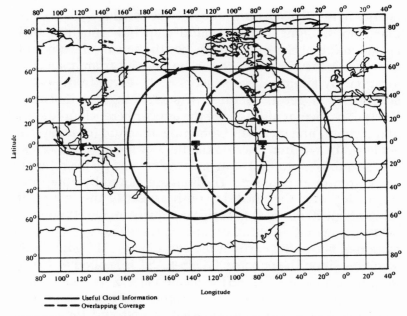

FIGURE 9. The field of view from the GOES-WEST and the GOES-EAST satellites. (Courtesy of NOAA.)

between visible and infrared. At nighttime, of course, visible images don't provide any useful information, but the infrared sensor is still capable of following the storm systems as they develop and propagate across the country. An example of a GOES full-disk image is shown in Figure 10.

At one time, NOAA would routinely generate a series of images composed of sectors of the full-disk image. This allowed meteorologists to concentrate on one particular portion of the earth. The most common sector shown is that of the entire United States. In addition, NOAA would rou-

FIGURE 10. The view of the earth from a GOES perspective. This type of image is called a full-disk image.

tinely make available even smaller sectors (seven in all) that gave the forecaster the ability to use 1-km resolution. The high-resolution images are extremely beneficial for forecasting severe thunderstorm development and for providing up-to-the-minute observations of approaching weather systems. With the availability of computer work stations at most National Weather Service forecast offices, each forecaster now has the ability to zoom in on particular regions of the atmosphere. The newer computers and their ability to manipulate vast amounts of data have made the NOAA service of providing predetermined sectors no longer necessary. Now all forecasters can manipulate the images and "see" what they want from space that applies to their own region.

On board those GOES satellites launched since the late 1970s is an instrument called the Visible and Infrared Spin Scan Radiometer (VISSR) Atmospheric Sounder (VAS). Data from the VAS can be used in conjunction with other known atmospheric properties to calculate atmospheric temperature profiles over a selected geographic region. The data are then sent to ground stations where computers collate and evaluate the data.

The VAS instrument has the capability of taking a new sounding every 30 minutes. During periods set aside for obtaining special data for experimental purposes, the VAS instrument can be programmed for a limited coverage mode. Selected latitudinal bands of varying north–south dimensions can be imaged to obtain information for calculating a vertical profile of atmospheric temperature and moisture. The VAS system can provide detailed information in the visible portion of the electromagnetic spectrum with slightly better than a 1-km resolution. Using its infrared sensors, VAS can provide data on the thermal properties of the atmosphere at a 7-km resolution.

Currently, VAS gathers information in 12 specific wavelength regions of the atmosphere. Within the next ten years, however, we will be able to improve our ability to "see" through the atmosphere. The next GOES satellite, for example, will have on board an improved version of the VAS that uses information from 18 wavelengths to determine the vertical distribution of atmospheric temperature. Because of the giant technological steps involved in the development of the geostationary satellite, the entire program has run embarrassingly behind schedule. The first of the GOES-NEXT series of satellites was originally scheduled for launch in 1989. These next five satellites are designated GOES-I, J, K, L, and M. (Satellites are commonly designated by letters before they are launched, and by numbers after they start sending data back to earth.)

GOES-8 (formerly GOES-I) was launched in April 1994. Images from the GOES-I through M satellites will be produced in five spectral channels instead of the two that are currently done routinely. Some idea of what kind of additional information will be available can be gleaned from looking at the images in Figure 11 obtained through the three sensors aboard one of the METEOSAT geostationary satellites. The left view is similar to what can be seen with the naked eye. This image is probably what the TIROS-1 satellite, launched more than three decades ago, would have sent back if the television cameras aboard it could have reached the geostationary altitude of 22,300 miles. From what we can see in the visible spectrum (the left image in Figure 11), clouds are white. The center panel shows the same view from a different perspective, using the infrared part of the spectrum. Some of the very bright clouds appearing in the image from the visible part of the spectrum are now barely discernible; this is particularly true for the clouds off the west coast of Africa.

The METEOSAT is the European version of GOES in the United States. And just as the British drive on the left side of the road, the convention for the METEOSAT images is that very cold temperatures in the infrared image (in the center) appear as black splotches. (We will see in Figure 12—

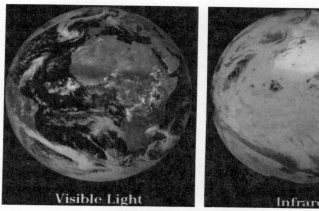

FIGURE 11. The earth as seen from METEOSAT. The visible image on the left is what is seen from the perspective of a satellite 22,000 miles above a point where the equator and Greenwich meridian intersect. The right panel shows an infrared image from a sensor that is sensitive to heat whereas the image in the lower panel shows an image from a sensor that is sensitive to infrared radiation emitted by water vapor molecules in the upper atmosphere. (Courtesy of D. R. Cahoon.)

visible and infrared images from GOES—that cold temperatures appear as white.) These black spots are most pronounced over eastern Africa and indicate the presence of tall
cumulus clouds producing tropical thunderstorms. We also
see a black swath associated with the clouds seen in the
lower left portion of the disks in these two images (extending southeastward from the east coast of Brazil). On the
other hand, some of the clouds that can barely be seen in the
visible image over central Africa now appear quite dark in
the infrared image.

The lower panel in Figure 11 shows an image from a
sensor aboard the METEOSAT that is sensitive to another
region of the infrared portion of the electromagnetic spectrum. The information contained in this image provides the
distribution of the amount of water in a layer of the atmosphere centered near 25,000 feet. In this image, areas containing a lot of water appear black. The black spots over
eastern Africa in the center panel also appear as black spots
in the water vapor image, confirming the presence of tall
rain-producing clouds. The relatively dark patches seen to
the west in the middle panel cannot be seen in the water
vapor image (except for two small dots along the western
Africa coast). The water vapor image at middle latitudes in
both the northern and southern hemispheres shows the
giant swirls of water vapor (darker areas) that are associated
with large middle latitude storms (also known as cyclones).
In addition to the dark swirls, we also see light swirls,
indicating very dry air. From a ground perspective, the light
areas would be the cold, dry air behind the front, whereas
the dark areas would indicate the presence of clouds and
precipitation ahead of the front.

Figure 12 compares the images derived from the visible
and the infrared sensors aboard the GOES-WEST satellite.

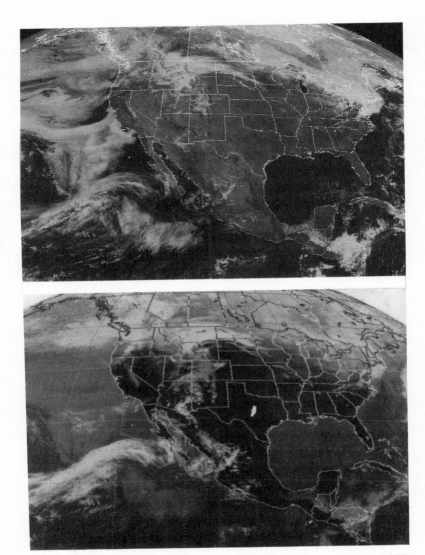

FIGURE 12. Visible (*top*) and infrared (*bottom*) images of the U.S. from GOES-WEST. These images show how stratus and cirrus clouds can be distinguished from each other. The bright clouds off the California coast in the visible image are a combination of low stratus clouds and higher clouds. The stratus clouds are not bright in the infrared image whereas the band of precipitation-producing clouds are bright in the infrared image. Many of the bright clouds over northwest Mexico in the infrared image cannot be seen in the visible image, a signature of wispy cirrus clouds in the upper atmosphere.

The visible image (top) is what your eye would see if you were looking at North America from a point 22,300 miles above the earth at a location on the equator. On this particular day, you would see a lot of cloudiness off the California coast. What you could not determine, however, are the kinds of clouds or the altitude at which these clouds were located. With the information provided by the infrared sensor, however, the different kinds of clouds in the eastern Pacific Ocean can readily be distinguished. In the infrared image, very bright objects indicate that we are seeing things that are very cold. Off the west coast of California and Mexico, most of the clouds are not very bright in infrared, which means that most of the tops of the clouds are not very cold. Because the atmosphere gets colder as the altitude increases, we know that most of the clouds do not extend very high and that the light gray areas in the infrared image are stratus clouds, and possibly even foggy areas near the coast. On the other hand, there is one line of clouds off the coast of the Baja peninsula (generally oriented in an east–west direction) that is also bright in the infrared image. These clouds extend much higher into the atmosphere and are very likely associated with the recent passage of a cold front. It is likely that this line of clouds also is associated with precipitation.

The infrared sensor is actually measuring what scientists call *radiance*. Radiance measurements can be calibrated to reveal an object's temperature. By determining the temperature, we can tell exactly how high the tops of these clouds are. Such information is particularly important for pilots, who thus can decide whether they can fly over the clouds or whether they should fly around this area of "weather" altogether.

Due east of this cloud band in the infrared image, we note some widespread cloudiness over northwestern Mexico. If, however, we look at the same region in the visible image, the presence of clouds is not obvious. Because of the latter fact, scientists call them "optically thin." Since, however, they do appear to be fairly bright in the infrared image, we know that they must be cold. Considering all of these clues, meteorologists can conclude that these cold but not visible clouds are cirrus. Composed of small ice particles, most cirrus clouds are located at altitudes of 30,000 feet or higher. They are the high, wispy clouds that mute the sun but don't cause precipitation.

Another curious feature of the infrared image is that the land over most of the United States and Mexico is dark compared with the oceans. This means that the land surface is warmer than the adjacent oceans, which is not surprising since this April image was obtained near local noon. The darkness indicates that the land masses have already absorbed some heat from the sun. In contrast, the northern United States and Canada show up as much colder in the infrared image. Looking at the visible image, we see that this region is also quite bright. Is the brightness caused by clouds or by snow? Considering only the GOES visible and infrared images, it is not an easy task to distinguish between regions of snow and regions of low clouds. Other information is required.

The spectacular extent of Hurricane Allen in 1980 is captured by the images shown in Figure 13. Allen was perhaps the largest hurricane ever captured by satellite imagery; the infrared image in the top half of the figure shows that its circulation was sucking in air even east of Florida. These high-altitude, cold cirrus clouds were not discernible in the visible picture. On the other hand, there is a bright visible region in western Pennsylvania (enclosed by the circle) that barely shows up in the infrared image. This turned out to be an area of low-level clouds and fog.

FIGURE 13. Infrared and visible images of Hurricane Allen, August 8, 1980. 16 GMT. *Top*: infrared image; *bottom*: visible image.

Without the two sensors aboard the GOES, this bright patch of clouds could have been mistaken for strong thunderstorms.

On the polar-orbiting NOAA satellites, meteorologists can make use of data from five channels. These additional channels provide data that are particularly useful for monitoring snow cover, tracking large ice floes, and even for monitoring widespread burning and clearing of forests and grasslands in tropical regions. Using this information, scientists can actually produce maps of fires started by farmers in Brazil, where it is now against the law to burn. Maps similar to those in Figure 14 can be produced shortly after the satellite has passed overhead, allowing government officials to catch violators as their fields burn and fine them on the spot.

Since polar-orbiting satellites pass over a given location only once or twice each day, data from them are most frequently used for research purposes rather than for real-time forecasting purposes. Information from the polar-orbiting satellites is entered into numerical models and used as an adjunct to provide forecasts. These data are called "asynoptic" since, unlike the synoptic soundings, all of which occur at the same time around the globe, they must be added some time after the computers have begun their calculations to provide forecasts.

In addition, since there are several NOAA polar-orbiting satellites normally in operation, these lower-orbiting, higher-resolution satellites do provide important data over a given area when forecasters are lucky enough to have them directed at the region they are interested in. Satellite receivers can even be put aboard research airplanes and, since at least some information is always available over any part of the globe, satellite overviews can be obtained several times each day from these portable receivers.

Although there are as yet no actual data from the new GOES-NEXT satellites, the Weather Service has already tested operationally how they plan to use such information. The GOES-NEXT series of satellites is much more complex than the current generation of satellites. For one thing, it will require the processing of information from 18 regions (or

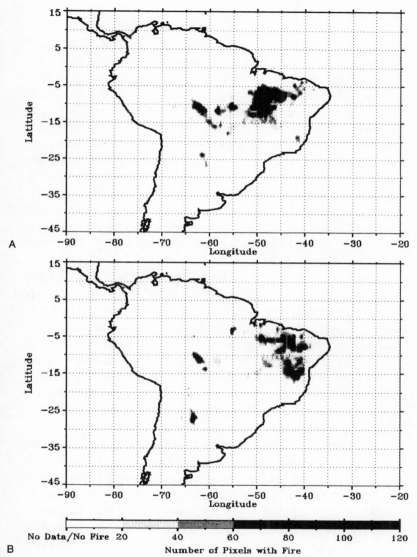

FIGURE 14. Satellite identification of fires in Brazil. The panels show the location of fires in Brazil derived from measurements on one of the polar-orbiting NOAA satellites in 1992. *Top*: August 26–September 3; *bottom*: September 25–October 1. The fire distribution makes use of special sensors that can uniquely determine the temperature of the ground. The two images show how location of the burning has migrated during a four-week period. AVHRR-based pixel fire counts; 1.0 × 0.5 degree bins. (Courtesy of V. G. Brackett and A. Setzer.)

channels) of the electromagnetic spectrum. With that many channels, forecasters will be able to receive both temperature and moisture profiles throughout most of the atmosphere.

What this means is that when these new satellites are operating, forecasters will have temperature and moisture data from areas where previously we had little, such as over the oceans. This information can then be fed into the computer models that crank out our daily weather forecasts. With no clouds to impede them, the new satellite instruments will be able to determine the temperature and moisture at 40 levels in the atmosphere (although moisture levels are available only below 30,000 feet; above that the air is too dry and cannot be measured accurately).

With such information, forecasters will be able to follow more closely exactly how weather develops. Much of our weather, both here and overseas, originates over the oceans. With the new generation of satellites and sensors, forecasters will be able to track air masses with more accuracy as they travel from ocean to land, seeing for the first time previously unknown details in the atmosphere over the remote mid-Pacific and mid-Atlantic. This kind of information can bring the "art" of weather forecasting closer to science.

But there is always a problem that affects measurements from space—clouds. Even the new generation of sensors are unable to penetrate cloud cover. Since the early 1980s, scientists have attempted to simulate how various kinds of clouds interfere with what the satellite sees, what effects clouds have on the data tranmitted back to earth, and what is actually present in the atmosphere. Algorithms have been developed that forecasters hope will help the sensors work around the presence of clouds. But until the GOES-NEXT generation of satellites is providing measurements, scientists won't know exactly how well their efforts have worked.

As exciting as the GOES-NEXT satellites will be, there are technological advances already on the horizon that may make the GOES-NEXT images as outdated as a black-and-white television set is now. Until now, we've been discussing what scientists call passive sensors. These are sensors that passively take readings of the atmosphere, using what is reflected back to them.

Scientists are currently developing an entire new generation of satellite instruments called *active remote sensors*. These will use laser sensors to probe the structure of the earth's atmosphere with more accuracy and with higher resolution than ever before. Once data from these instruments become commonplace sometime in the next century, it is possible that the remote sensing which we currently view as so impressive will have suffered the same fate as the 78-rpm record.

A STEP INTO THE FUTURE: THE EMERGENCE OF THE LASER AND ATMOSPHERIC MEASUREMENTS

Soon after the development of the pulsed ruby laser in 1960, scientists used it to study the atmosphere and to measure distances accurately in space. The pulsed laser sends out an intense beam of light at exactly one wavelength. What makes the laser so powerful is that its light is coherent, i.e., does not spread out like the beam of a flashlight.

The development of laser technology, including lidar (*l*ight *d*etection *a*nd *r*anging), has made possible an entirely new way of measuring what is present in the atmosphere.

Called *active remote sensing*, most people are already familiar with one common type already in widespread use, i.e., radar (an acronym for radio detecting and ranging). The radar sends out a beam of electromagnetic radiation (in the microwave portion of the electromagnetic spectrum) and measures the return signal as a function of time. The time it takes for the reflected waves to return to the receiver determines its range. If no signal is returned, then the beam has not been intercepted by anything. But if something is present in the field in front of the transmitter, then a signal can be picked up by a receiver that is co-located with the transmitter of the outgoing signal. The longer it takes for the signal to reach the receiver, the farther away the object that intercepted the radiation. By knowing the time history of the reflected beam of radiation, we can determine how far away these reflecting particles were and then derive a picture of the distribution of the reflecting particles, which are mostly water droplets and ice crystals embedded within a cloud. Larger pieces of ice, especially hail, are very efficient reflectors and a very strong return is linked with the presence of hail, which is often associated with severe weather.

Another form of radar uses a laser to send out pulses of energy at a precise wavelength so that the amount of the signal returning to the receiver can also be translated into information as to what is present in the atmosphere. In its simplest form, this laser–radar system fires a beam of laser light through the atmosphere and any particle in its path scatters some of the coherent beam. As the light hits these particles, some of it bounces back to a receiver mounted near the source of light. Used in this manner, the lidar system can "see" very thin layers of haze or particles, allowing it to detect regions in which clouds will form before the clouds coalesce into large enough particles that can be seen by the naked eye.

Scientists began using lasers to measure atmospheric distances within a few years of their development. As early as 1964, measurements of distance to the Beacon–Explorer series of satellites were routinely made with a high-power, long-pulse, low-repetition-rate ruby laser. These early distance measurements achieved an accuracy of several meters. The following years produced a rapid improvement in ranging precision to the current accuracy of less than one centimeter.

During this same period, scientists recognized the potential of the laser ranging capability to observe the atmosphere. The earliest atmospheric lidar measurements were used to determine the distance to clouds and aerosol layers. Some of the first experimental lidar atmospheric experiments were conducted to detect meteoric and volcanic debris in the stratosphere.

A result of one of these experiments is shown in Figure 15. A fire had been burning in Yellowstone National Park for several weeks. In Figure 15, the set of measurements, obtained from a ground-based lidar system, depicts the vertical distribution of aerosols (particles in the atmosphere) from the fire. The lidar system that obtained these particles was located at NASA Langley Research Center in Hampton, Virginia, more than 2000 miles away. What can be seen in this figure are two distinct layers of smoke at 5 and 7 kilometers. These layers, as well as the layer of particles at 11 km, which are microscopic particles that will later become the condensation nuclei for cirrus clouds, were not visible. In other words, the lidar was able to see exactly where clouds would form before they could ever be seen.

Perhaps the most difficult task for forecasters currently

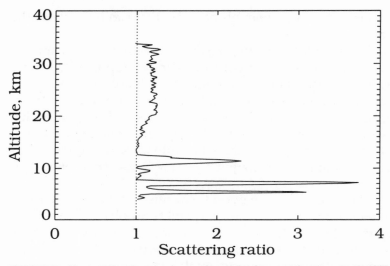

FIGURE 15. Vertical distribution of aerosol particles from smoke, August 23, 1988. The lidar system at NASA Langley Research Center in Hampton, Virginia, shows the presence of two layers of smoke at 5 and 7 kilometers. This smoke has been transported more than 2000 miles from fires burning in Yellowstone National Park. An enhanced scattering ratio, a measure of how much light from the lidar is reflected back to the ground, is also observed at 11 km and indicates the presence of microscopic particles that will eventually form cirrus clouds. (Courtesy of M. Osborne and M. P. McCormick.)

is to predict the formation of clouds. Imagine, however, if we had satellites orbiting the earth that were able to pinpoint the preferred locations of clouds because they could "see"— in advance—where the clouds would be hours before they were actually there. This dream will become a reality when such lidar systems are orbiting the earth routinely.

Another intriguing set of lidar data from the Langley Research Center are shown in Figure 16. This particular depiction shows the impact of Mt. Pinatubo on the distribution of aerosol particles in the stratosphere. Mt. Pinatubo erupted on June 11, 1991, and by August 28, the cloud of sulfuric acid particles had reached the northern middle latitudes. Although these signals are unmistakable in the lidar measurements, they could not be seen by the naked eye. It was these particles that gave us our spectacular sunsets in 1991 and 1992. If we had a lidar capability from a space platform that routinely circled the globe (instead of on the ground looking up, which is how the measurements in Figure 16 were obtained), we could determine the circulation patterns in the stratosphere much more accurately than can be done by any method presently available. Such information could possibly take us another quantum step forward in our ability to predict the weather.

The lidar system at NASA Langley has been making continuous measurements since 1974. The particles that give rise to the layers shown in Figures 15 and 16 act as tiny little mirrors that reflect back to a receiver on the ground. A measure of how much light is bounced back to the ground is called the *scattering ratio* and we can see this quantity plotted as a function of altitude.

The total amount of light scattered by all the particles in the atmosphere can be quantified by a term defined as the *integrated backscatter*. The integrated backscatter measured by the NASA Langley ground-based lidar system since 1974 is shown in Figure 17. Note that the scale on the left side of the figure is logarithmic. Along the bottom axis are arrows which point to instances when a volcanic eruption took place. This figure illustrates the magnitude of the El Chichón eruption in Mexico in 1982 and the Mt. Pinatubo eruption in the Philippines in 1991 relative to other volcanoes.

To see whether or not a lidar system can work in space, a project called LITE (*Laser-in-*Space *Technology Experi-*

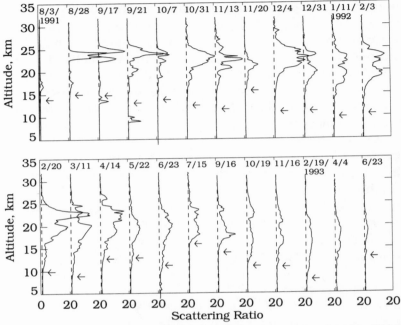

FIGURE 16. A time line history of the impact of Mt. Pinatubo. A series of lidar profiles from the NASA Langley lidar between August 1991 and June 1993 is shown. Layers of particles thrown into the atmosphere from the eruption of Mt. Pinatubo in the Philippines in June 1991 show up in the stratosphere in layers generally located near 20 km (~70,000 feet). The arrows in each of the profiles show the height of the tropopause, the boundary between the troposphere and stratosphere. (Courtesy of M. Osborne and M. P. McCormick.)

ment) is scheduled to be flown aboard the Space Shuttle in late 1994. LITE will be the first lidar in space. Somewhat analogous to the simplistic goal of TIROS-1 more than 30 years earlier, the purpose of LITE is primarily to show that useful data can be obtained from a lidar in space. Scientists hope to obtain from LITE a global snapshot of haze and clouds.[14]

The laser aboard LITE will fire three laser beams toward earth at three precise wavelengths: 1.06 micron (a micron is one millionth of a meter), 0.53 micron, and 0.355 micron. The first wavelength is in the infrared part of the spectrum and the latter two, the visible part. By looking at the return signals from these three separate wavelengths, scientists can gather important information about the size of the particles in the atmosphere, since particles of different size reflect light in a different manner. (The size of the particles and

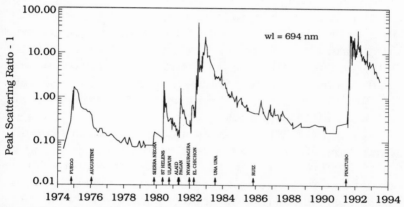

FIGURE 17. Time line of lidar measurements. Peak aerosol mixing ratio: aerosol lidar measurements at LaRC (37°N, 76°W). The impact of volcanic activity is evident from this time line of integrated backscatter, a measure of how much light is reflected back to the ground by aerosol particles. The arrows along the time line indicate when eruptions of various volcanoes occurred. The scale on the vertical axis is logarithmic and indicates that the backscatter following the Mt. Pinatubo eruption resulted in an enhancement in backscatter more than 200 times greater than when the atmosphere was relatively devoid of volcanic debris in the late 1970s. (Courtesy of M. Osborne and M. P. McCormick.)

water droplets in a cloud is one of the critical factors which determine whether a cloud appears fluffy white, gray, or even black.) The vertical resolution of the data should be better than in any previous attempt to see particles and clouds in the atmosphere using passive remote sensing techniques.

But the lidar system aboard LITE is already quite primitive compared with other lidar systems that have been used to study the atmosphere since the late 1970s. A major advance in the development of lidar systems is the *Differential Absorption Lidar* (DIAL), which can actually provide vertical profiles of trace gases in the atmosphere.[15] The DIAL system fires two laser beams simultaneously and takes advantage of one of the fundamental laws of quantum mechanics. According to the quantum theory, every molecule absorbs and emits energy only at specific wavelengths. Two molecules that atmospheric scientists have been most interested in over the past few decades have been ozone and water vapor. One of the advantages of lasers is that they can be "tuned" exactly to a frequency that is known to be absorbed by that molecule.

So if, for example, water vapor absorbs radiation with exactly the energy characteristic of 0.72043 micron and a lidar signal is sent out at that wavelength, the return signal would be diminished by both the amount of water vapor in the atmosphere and whatever other particles are in the path of the lidar beam. If at the same time we send out another beam at, say, 0.72100 micron (which would not be sensitive to water vapor), we would have two readings to compare. The difference in the return of the two signals must be related solely to the absorption of the beam that is sensitive to the presence of water. By looking at the range-resolved *difference* between the two signals, the distribution of water

vapor can be determined. To date, such a system has operated successfully from the ground (looking up) and from an
airplane (looking down and/or up).

The next major test of the DIAL system is a program
called LASE (*Lidar Atmospheric Sensing Experiment*). The
purpose of LASE is to fly a water vapor DIAL system on an
experimental ER-2 airplane. The ER-2 is a converted U-2 spy
plane that flies in the stratosphere (65,000–70,000 feet, or
19–20 km) and has been modified to use instrumentation for
environmental research. Another goal of the LASE project is
to use a system as complicated as the water vapor DIAL in a
completely "hands-off" mode. This first flight of LASE is
scheduled to take place in the mid-1990s, when scientists
hope to determine the feasibility of using differential absorption lidar systems in space.

Another important instrument of the future is the Laser
Atmospheric Wind Sounder (LAWS). This type of laser
system uses Doppler lidar measurements to derive wind
speeds. A Doppler system is similar to that used in radar
guns to measure the speed of anything that moves, from
speeding automobiles to a tennis player's serve.

In general, Doppler radars work on the principle that
after a beam of energy hits a moving object, the reflected
energy returns at a slightly different wavelength. Doppler
radar detects the difference between the frequency at which
the beam is emitted and the return frequency. One of the
most common examples of the Doppler effect is the sound
you hear when a train rumbles by. As the train approaches,
the sound that it makes is determined both by the actual
noise of the train and the fact that the sound waves reaching
your ears are coming toward you faster than the speed of
sound because they are also traveling with the speed of the
train toward you, much like a speedboat pushes a wave

ahead of its bow. As the train passes, it makes the same noise, but now the sound waves are traveling away from you at the speed of the moving train. As a result, you hear the train coming toward you at a higher frequency than when it moves away from you. The difference between the frequencies of the approaching and retreating sound waves can be measured to determine the speed of the train.

Another common example is the radar gun which fires a beam of waves at a moving object to determine its speed. The frequency of the emitted wave is known and that of a wave bouncing off the moving object (e.g., a car, or a baseball) can be measured to determine the speed of the moving object. The principle of Doppler radar has now become more commonplace in weather forecasting, and many regions around the country display information from local Doppler radars as part of their routine weather reports. These radars not only show where rain is located within a cloud but also can determine the velocity at which these particles are moving, again indicating the intensity of the storm by measuring the velocity of the wind circulation. The magnitude of the difference is directly dependent on the speed of the moving object. If there are particles (such as dust) in the atmosphere (and there always are), then the speed of the wind can be obtained.

Thus, the discipline of satellite meteorology will advance significantly in the near future, first with the launch of the GOES-NEXT series of satellites and then with the development of lasers in space that will be able to obtain information about the atmosphere in such detail that we don't even know how it might best be utilized yet. The data from the GOES-NEXT satellites over the next one to two decades should eventually result in formidable gains for the numerical forecast models that are currently in use. The incorporation of these data for forecasting promises to be one of the most intense research areas for the first part of the 21st century.

The lidars in space, however, are still in the developmental stage, just as the first TIROS satellites were only in the conceptual stage in the 1950s, even though weather forecasters knew then that pictures from space would be an important addition to weather forecasting. Will systems like LITE and LASE be as commonplace by the middle of the 21st century as GOES and polar-orbiting satellites are today? At this point, it's hard to know for sure. But then again, did anyone envision the compact disk when high-fidelity 33-rpm records became the standard only 40 years ago?

FIVE

The Science of Hurricane Forecasting: Saving Lives Is the Bottom Line

The plane, a P-3 Orion, is ready to go. It sits on the runway in south Florida, not far from the Hurricane Forecast Center of NOAA in Miami. The day is typical of southern Florida in September—hot, humid, with hazy sunshine. Not much happening. But a couple of hundred miles out to sea there is a tropical storm that is about to be upgraded to Hurricane status by the National Weather Service. Though the storm is not about to come ashore anywhere and poses no threat to any population, the National Weather Service is about to track it. To track it correctly, they must find out exactly where the eye of the storm is.

The P-3 Orion is the same kind of plane the Navy used to use to track Soviet submarines during the Cold War—a big, lumbering workhorse of a plane. Inside this one the bulkheads are lined with dials and switches and instrument readouts. These will be read by a team of meteorologists and technicians, who will then direct the pilot of the plane so he can fly the plane directly into the eye. The process is somewhat like the blind man with good legs leading the lame man with good eyes, a symbiosis between the scientists and pilot.

The plane takes off and heads toward the storm. It's in the air for over an hour before the members of the crew begin to see their destination. On the horizon is a line of dark, ominous clouds. These are the spiral bands, the edge of the hurricane. Inside the plane the instruments are showing a drop in air pressure. Radar shows the spiral bands as bright red and yellow, indicating intense thunderstorm cells. Suddenly the plane is flying through them, winds buffeting the fuselage, the passengers inside experiencing the same kind of ride as they would sitting in the back of a truck going over a bumpy road—only they are 8,000 feet above the ocean, suspended in the middle of one of the most destructive forces in nature, with nothing between them and annihilation but an inch or so of aluminum and a lot of luck.

Little light enters through the portholes because outside it is twilight, a gloomy grayness broken only by the rain splattering against the window. There is nothing to see outside, no telling which way is up, which way down; the plane is encased in a soft, gray, wet world.

But there are other ways of "seeing," and inside the plane the scientists study the dials of their machines and instruments, oblivious to the howling winds and lightning and torrential rains they are flying through. Barking suggestions to the pilot, they help steer the plane to the very center of the storm, the "eye," getting their leads from the gauges recording the dropping pressure and the shifting winds. Knowing how the winds circulate inside a hurricane, they are able eventually to steer the plane into the calm center, where the pressure has dropped to its lowest.

Outside the plane's windows is a different world. Above the sky is blue, below the ocean water is devoid of whitecaps. Yet on the horizon is an impressive wall of dark clouds, the eyewall of the hurricane. The gauges affirm what the eyes see—the pressure has remained steady, the wind has died down to below 10 mph. This is the dead calm of the storm, the "eye." One of the scientists charts the plane's location, then radios the storm's location back to the hurricane forecast center. Their mission has been successful, they have found the eye and plotted it on the map so it now can be tracked.

Suddenly the windows darken, the plane begins bouncing, the gauges go wild. Wind speeds suddenly jump to 50, 60 mph. The plane is piercing through the other side of the hurricane. The only difference here is the direction of the winds as they rotate around the eye. Finally they are out, the grayness outside lightens, the air pressure rises, the winds drop off. They have flown through a hurricane.

On September 16, 1938, a Brazilian freighter was en route to New York when it sighted a dangerous tropical storm about 350 miles northeast of Puerto Rico. In Jacksonville, Florida, Weather Service meteorologists took note of the storm. It appeared to be heading directly for Miami. By the next morning the hurricane had moved 250 miles closer, on a direct bead for Miami. The Weather Bureau issued an urgent warning to the citizens in the Miami area that the hurricane would reach them within 24 to 36 hours.

In Miami, accustomed to hurricanes, the residents began boarding up windows and taking to the high ground. By next morning, however, they received good news from the Weather Bureau: the storm had slowed down and changed its course, it was now heading northeast and was expected to turn out to sea past Cape Hatteras, North Carolina.

But this was 1938. Radar was still in its infancy. Observing stations were limited to those on land and ships at sea. The Weather Bureau in those days relied on voluntary reports from merchant ships and commercial aircraft for news of ocean weather. There were no satellites in the sky beaming back weather maps every four hours. There weren't even that many planes flying across the ocean. What this meant was that there was no way of knowing what the hurricane was up to. Had there been a weather map of the Atlantic Ocean and the eastern seaboard of the United States available that week, it would have shown a high-pressure area over the Midwest and the Alleghenies, and another large high-pressure cell drifting down from Nova Scotia to the Bermuda high. The hurricane was situated in the valley formed by these two high-pressure areas, a valley leading right up the East Coast to Long Island, an expressway of destruction.

Near Cape Hatteras the storm brushed the Cunard White Star liner Carinthia, southbound with 600 passengers on a cruise from New York to the Caribbean. The ship's captain had responded to the Weather Bureau's prediction of the storm turning northeast by changing his course to move closer to Cape Hatteras. But on the morning of September 21 his ship was being buffeted by the western edge of the storm. The ship's barometer read 27.85 inches, one of the lowest barometric readings ever recorded on the Atlantic coast. The storm was big and powerful and obviously was not heading out to sea.

Meanwhile on Long Island and in New England, the weather seemed balmy and warm. It was the equinox, when the sun and moon exert their most powerful influence on tides, and tides in the area were expected to be higher than normal even without the hurricane. After the report from the Carinthia, the Weather Bureau didn't receive any information at all on the storm for the next six hours. The Weather Bureau had no way of knowing that the storm's movement had increased substantially, traveling at the unheard-of speed of 60 mph.

Having no information, the Weather Bureau in Washington, D.C., which would have jurisdiction over the hurricane after it passed Jacksonville, issued no warnings to the East Coast to prepare for a storm. In fact, the forecast for September 21 for the eastern New York–New England area called for "rain, heavy at times, today and tomorrow. Turning cooler." In *The New York Times* that morning was an editorial complimenting the Weather Bureau for its alertness in warning Florida of the hurricane that was now "safely out to sea."

It wasn't until 2 P.M., one hour before the storm smashed into Long Island's south shore, that the Weather Bureau issued a gale warning and mentioned that a "tropical storm" would pass over the area. Even then, they didn't seem to realize the intensity of the storm, which would eventually kill 600 people, cause massive forest blowdowns, severe flooding, and result in more than $300 million in damage. Historians would call it the Great New England Hurricane because of the massive destruction it wrought over the entire six-state area. No one was prepared for the hurricane, since there hadn't been a hurricane come ashore in that area since 1815.

The hurricane smashed into Long Island between the towns of Babylon and Patchogue on the south shore at 3 P.M. The storm surge that followed lifted entire houses off their foundations and swept them out to sea. The storm then traveled across Long Island Sound and bludgeoned the coast of Connecticut between the cities of New Haven and Bridgeport, where winds tore up miles of railroad track and flung railroad cars as if they were toys. Traveling north through the Naugatuck River valley, the howling winds of the storm blew down thousands of acres of forests, destroyed entire villages, and tore up bridges.

In Providence, Rhode Island, east of the storm and thus subject to the strongest winds, winds pushed a tidal wave before it that flooded the entire downtown area, bringing that city to a virtual halt. Most of the entire state was under water, and Rhode Island had a death toll of 380 people from the hurricane, over half of the storm's total. The massive storm surge that hit the coasts of Rhode Island and Massachusetts created waves as high as 40 feet, washing over everything in their path and changing the land configuration. Winds at the Blue Hill Observatory in Massachusetts were clocked at 121 mph. Not only were the larger cities of New England affected, but all major rivers in New England reached flood stage because of the massive rainfall. The eye of the storm tracked across the Berkshires of western Massachusetts and through the center of the Green Mountains of Vermont before finally running out of steam as it traversed Lake Champlain and into Canada.

It is difficult to imagine today a scenario like that of September 21, 1938. Meteorologists can now spot hurricanes as they form off the coast of West Africa, thanks to the satellite images beamed down regularly from geostationary satellites. They can send hurricane hunters out to gather data about such storms long before the storms are in a position to cause damage to populations. The National Weather Service is able to issue hurricane watches and warnings days in advance of a storm's arrival, so that even when powerful storms do come ashore in populated areas, such as Hurricane Andrew in 1992, the loss of life is kept remarkably low. We have made progress in controlling the damage wrought by hurricanes, but there is still improvement to be made.

THE ORIGINS OF HURRICANES

The word *hurricane* comes from the Spanish, *huracán*, and is used to describe a tropical, rotating storm with winds above 75 mph. A hurricanes can be thought of as a mature form of a low-pressure cell, or cyclone, the kind of storm system we see on our nightly television weather report. In fact, hurricanes are called *cyclones* in the Indian Ocean. The difference between a "normal" cyclone and a hurricane is one of degree. Hurricanes are tropical rather than temperate in nature. Such storms are called *typhoons* in the Pacific. Hurricanes or typhoons are most often found in the southwest North Atlantic, the southeast North Pacific, and the southwest North Pacific, as well as in the Indian Ocean around Australia. No hurricane has ever been reported in the South Atlantic ocean.

The hurricanes that affect the United States originate over the warm waters off the coast of West Africa, usually along the Intertropical Convergence Zone (ITCZ). This is an area driven by the Hadley cell circulation discussed in Chapter 2, where the sun's energy heats vast reservoirs of water, thereby fueling the planet's "weather engine." The energy produces towering cumulus clouds often reaching 12 or more miles into the atmosphere. Some of the clouds become organized, forming clusters, and some of these clusters eventually develop circulations embedded in "waves" that propogate from east to west.

Hurricanes begin life as small cells of low pressure that

form on these waves at the edge of the ITCZ. After a small cell forms off the coast of Africa, it is pushed westward by the prevailing easterlies in the tropics. There's a lot of warm water between the coast of Africa and the islands in the Caribbean, which helps catapult what is at first merely a tropical storm into a full-fledged hurricane by the time it reaches the western hemisphere. As the winds around the cell increase, so does its official designation, from a tropical depression (wind speeds less than 39 mph) to a tropical storm (winds between 39 and 73 mph) to a hurricane (wind speeds over 74 mph).

Although destructive to human habitation, hurricanes are one of the most efficient heat transfer mediums in the world. The hurricane is an idealized microcosm of the global-scale "heat engine," possessing a central core warmer than the surrounding air temperatures at all levels. This warm core sucks up water vapor, which condenses into thick clouds, producing thunderstorms and even tornadoes at its outer fringes and heavy rains near the core. It is this intense condensation of water vapor that manages to transfer heat. So hurricanes are a means by which the atmosphere transfers heat from the tropics to the temperate zones, and a quite dramatic means at that.

Just what part do tropical storms and hurricanes play in the dynamic interchange of our atmosphere? From our previous discussion on the origins of weather we know that the sun shines more directly on the equator than on the poles. The angle of the sun becomes more oblique the higher the latitude. So we know that the tropics, the area geographically defined by the Tropic of Cancer and the Tropic of Capricorn at approximately 23° on either side of the equator, receive more solar heat than the rest of the earth. This is why the tropics are always warm, because the sun is more directly overhead all year long, passing directly overhead twice a year as the sun migrates from the northern tropics to the southern tropics.

The tropics are so warm you might say they have a heat asset, i.e., there is more heat coming in than is radiated out. The zones to the north and south of the tropics on up to the poles are areas where the heat budget is less, where there is a deficit. In these areas during certain parts of the year the net outflow of heat due to radiation and other causes is more than the inflow. Hurricanes almost always form in the area between these two zones, and slowly make their way north (or south in the southern hemisphere) with their tremendous store of heat energy. In other words, hurricanes act as a safety valve, a way for the atmosphere to let off a little steam.

Three ingredients are necessary for a hurricane to form. There must be heat, moisture, and unstable air. Stable air is air that is not prone to the formation of thunderstorms. Unstable air, on the other hand, has a vertical temperature structure such that when air from below is forced to a higher altitude by some external force, the air continues to rise because the air parcel arriving from below is warmer than the air of the new environment. Under such conditions, tall cumulus clouds are easily formed. The heat and moisture are easy to come by in the tropics, but instability is another matter. The high moisture content is easily obtained in the warm waters of the tropics, but there also must be enough instability at the upper levels of the atmosphere to allow rising warm air currents to ascend to considerable heights. The actual physical process by which this occurs is still unclear, but scientists do know that only about 10% of potential hurricane situations actually produce a storm.

What is needed to put it all in place is a "wave" action— a low-pressure trough in the upper atmosphere that moves west (in the North Atlantic hurricane breeding grounds) from West Africa to the eastern Caribbean. This wave is the trigger needed to disturb the normally homogeneous tropical air mass, enhancing rising convective currents that produce clouds and thunderstorms. The rising air creates even lower pressure at the surface, while the Coriolis effect sets up a rotational pattern. As convection increases, it sets in motion a chain reaction of more condensation, which results in lower pressure and more rising air and more circular winds. With a continued fall of pressure, the cyclonically turning winds accelerate and a central eye forms, extending from the surface of the sea to thousands of feet in elevation. Thus, a tropical storm is born.

Perhaps the most intriguing aspect of a hurricane is its eye. The center of a well-developed hurricane is an area of relative calm that rises like a column from the surface to the top of the storm structure. Sometimes there are a few clouds, but blue sky and calm winds are the distinguishing characteristics of a storm center. The diameter of an eye varies from storm to storm, but averages from 12 to 30 miles.

Meteorologists have made great strides in their understanding of hurricanes. Now that we can see hurricanes from satellites and track them with our radar and "hurricane hunter" aircraft, you would think we had solved the "hurricane problem." But it's not enough to know how strong the winds are in the hurricane, and how large the eye is and what the barometric pressure is. What's more important for the people who live in hurricane areas is this: Where is it going and when is it going to get there?

Determining where a hurricane is going is one of the trickiest parts of hurricane forecasting. A storm's track depends on several factors. Most importantly for North America is the status of the Azores–Bermuda high-pressure area.

This anticyclone sits in the southern North Atlantic like a mountain, directing weather around it. The hurricanes that affect the United States usually form in the ocean off the coast of West Africa and move westward on the prevailing easterly winds of the tropics. These winds generally take the storms through the West Indies and into the Caribbean Sea or the Gulf of Mexico. However, if the high pressure weakens a bit, and there is a low-pressure trough extending southward from the United States mainland, the storm can easily turn northward and eventually make landfall on the southeast coast of the United States. The exact track the storm takes, i.e., whether it makes landfall on the southeast coast or continues up the coast in a more northeasterly direction, depends upon the location and bearing of the low-pressure trough.

Even more difficult a task is predicting the exact time of landfall. Hurricanes travel at varying speeds, and the speed can change dramatically. For example, the Great New England Hurricane of 1938 picked up speed only after it had passed Cape Hatteras. Most storms travel at speeds of from 10 to 15 mph while they are still in the tropics, then pick up speed as they curve northward to catch a ride with the prevailing westerly winds. There are times when a hurricane grinds almost to a halt, a genuine catastrophe for those caught on the ground.

The basic structure and dynamics that we've just discussed in relation to the North Atlantic hurricanes apply to the Pacific typhoons and storms in the Indian Ocean also. In the Pacific the major storm-producing area is east of the Philippine Islands, and in the Indian Ocean it is in the Bay of Bengal. What is true about hurricane formation in the North Atlantic is true everywhere: hurricanes generally form in tropical waters along a "wave" where tropical and polar air masses converge. The world's largest "factory" of hurricanes is the area of the southwest North Pacific east of the Philippines, where an average of about 20 typhoons originate a year. The Atlantic breeding grounds, on the other hand, produce an average of seven storms per year.

Hurricanes are one of the most powerful forces in nature. The energy produced by a full-fledged hurricane is awesome. The lowest barometer reading ever recorded was taken in the midst of a hurricane. The S.S. Sapoerea was at sea east of Luzon, Philippine Islands, during a typhoon when its barometer recorded a reading of 26.185 inches of mercury. That was the record until an aircraft reconnaissance plane dropped a sounding in the eye of Super Typhoon Tip on October 12, 1979 and recorded a reading of 25.69 inches. That reading was taken very near the earlier record reading, between Luzon and Iwo Jima.

In the Atlantic, Hurricane Camille produced a barometric reading of 26.73 inches just before landfall in the Gulf of Mexico in August 1969. This extremely low barometric pressure at the eye of the storm is what produces such extraordinary winds. We know that winds are caused by air moving from high pressure to low pressure. With extremely low pressures the gradient is so steep that winds pick up extraordinary speed. If you can remember riding a sled down a hill, you will understand how the steepness of the hill causes the sled to go faster. Together with the centrifugal force, which whips the wind around the eye of the hurricane and pushes it even faster, this causes the highest wind velocities to occur in the upper right quadrant of the storm. Thus the damage the storm wreaks depends often on where you are in relation to the eye of the hurricane.

Although a hurricane's strong winds are one of its most damaging characteristics, they are not what causes the most damage in terms of money. Since a storm's strongest winds are confined to one quarter of its area, and since the winds die down rapidly once a hurricane moves onto land, a far more damaging characteristic is the *storm surge*. It is this aspect of storms that has caused the most damage in recent years, especially in areas near the sea where the population has increased. As a hurricane crosses the continental shelf, it pushes a tidal mass of water before it. Water levels can increase as much as 15 feet during a storm surge. The surge occurs before and slightly to the right of the path of the eye. In addition to this extraordinary tide, wind-whipped waves are added to the top of the tidal mass. They can cause extensive damage to coastal areas and offshore islands. Areas of Long Island and Cape Cod, for example, have been geographically reconfigured as a result of high water from hurricanes. In fact, damage from storm surges generally exceeds that from high winds, especially along the Gulf Coast and southeastern coast of the United States, where much of the land lies only about 10 feet above sea level.

In the Far East, storm surges kill millions when cyclones, as they are called in that area, come ashore along the Bay of Bengal. In 1970 a cyclone drowned 200,000 in what was then East Pakistan (now Bangladesh).

HISTORY OF HURRICANE FORECASTING

We've come a long way since the devastating hurricane of 1938, when the National Weather Service relied solely on ships and flying aircraft for information of weather over the oceans. Now, thanks to technological advances that allow us to measure and display weather information in real time on mass media such as the Weather Channel, we can track hurricanes and participate in them while safely ensconced in our homes before the television screen. We can fly through the eye of a hurricane, we can taunt the massive storm from ashore like a matador taunting a bull, confident that our satellites and radar will give us enough time to prepare for the storm's onslaught.

Obviously, hurricanes have been around for as long as civilization has inhabited coastal areas prone to such storms. In the western hemisphere, Christopher Columbus mentioned such storms in his diary. The ever-curious Ben Franklin studied hurricanes and was the first to discover that the storms moved from one place to another but were not steered by surface winds. But it wasn't until 1847 that a hurricane-warning system was first established in the United States. Lt. Col. William Reid, of the Royal Engineers of England, while on duty in Barbados, devised a system of hurricane warning based on barometric readings.[16]

A few decades later Father Benito Vines, director of Belen College in Havana, Cuba, developed a systematic means of forecasting hurricanes using observations of upper- and lower-level cloud formations. He routinely issued hurricane warnings beginning in the 1870s. At the same time in the United States, the government was appropriating money to set up a national meteorological service under the auspices of the Army's Signal Corps. The newly organized service recognized immediately the need to provide hurricane warnings for the eastern and southern coasts of the United States. To do this the meteorological service set up a system to receive daily observations from islands in the Caribbean.

Using this new observational network, the meteorological service issued its first hurricane warning on August 23, 1873, for New England and the Middle Atlantic states.

Although the storm wasn't a full-fledged hurricane, the warning was the first official warning of a storm of tropical origin made by the fledgling meteorological service.

In September 1875, a strong hurricane destroyed the entire town of Indianola, Texas, killing 176 people. The storm arrived unannounced, despite the warning system set up by the meteorological service. It was this catastrophe that sparked the eventual creation of the U.S. Weather Bureau in 1890 and its transfer to the Department of Agriculture.

The killer storm of 1875 that came ashore unexpectedly illustrated the weaknesses of the fledgling hurricane warning system. Part of the problem at the time was that hurricanes were not perceived as that much of a threat to the general population. There were relatively few people living along the Florida coast then. The cities of Miami and Fort Lauderdale were still mostly undeveloped swampland. Along the Gulf Coast there were larger population centers, cities like New Orleans and Mobile, but these areas were still perceived as "out west" to the major population centers in the Northeast.

It took the Spanish–American War at the close of the 19th century to revive the urgency aimed at establishing a comprehensive hurricane warning service for North America. After President McKinley declared that he feared a hurricane more than the Spanish Navy, the U.S. Congress authorized funds to establish observation stations in the islands of the Caribbean. Such stations were set up on the islands of Jamaica, Trinidad, Curaçao, Santo Domingo, and Cuba. Although their original purpose was military—to warn the U.S. troops in the area of approaching storms—following the war's end the observing stations became a permanent part of the nation's hurricane warning system.

With the addition of observing stations on Puerto Rico and Dominica, the headquarters of the system was moved from Jamaica to Havana, Cuba.

The services of the storm warning center were made available to all shipping and commerce in the area. But despite the observing stations, the government maintained the practice of issuing hurricane warnings from its Washington, D.C. headquarters. It was a catastrophic decision, since communication in those days between the Caribbean and Washington was primitive and slow. On September 9, 1900, the most destructive hurricane ever to hit the United States came ashore at Galveston, Texas, killing more than 6,000 people. Galveston sits on a sandbar in the Gulf of Mexico, and the storm surge from the hurricane washed completely over the island, resulting in massive destruction and loss of life. There is no record of any warning ever reaching the area.

The catastrophe in Texas forced the Weather Bureau to move the hurricane forecasting headquarters from Cuba to Washington, in the hope of improving service to those areas on the continent prone to hurricane visits. But this move simply exacerbated the problem, creating a situation in which the people most affected by hurricanes were dependent on warnings from an area seldom affected by the storms. This problem came to light most dramatically in 1926 when a destructive hurricane bore down on the fledgling town of Miami and the Weather Bureau in Washington didn't get around to issuing a warning until 11 P.M., after most of the residents had gone to sleep. Two hours after the hurricane warning was issued, at 1 A.M., hurricane-force winds hit the city, causing massive destruction.

By 1935, such incidents led Congress to appropriate $80,000 to revamp the hurricane warning service. This influx of funds allowed the Weather Bureau to set up forecasting centers in Jacksonville, New Orleans, San Juan, and Boston. Each of the forecasting centers would have its own area of responsibility: San Juan would be responsible for the Caribbean Sea and islands east of longitude 75° W and south of latitude 20° N; New Orleans for that portion of the Gulf of Mexico and its coasts west of 85° W; Jacksonville would cover the remaining portions of the Atlantic, Caribbean, and Gulf areas south of 35° N; and Washington and Boston would share the area north of 35° N.

Each of the forecast centers would issue hurricane advisories four times a day. On paper, the new organization looked good, but it was woefully understaffed. The Jacksonville center, for example, had only two forecasters working 24 hours a day, seven days a week. This system continued for years. Although it was an improvement over the previous system, it was still too easy for storms out at sea to fall through the cracks of the Weather Bureau's ability to track them, as was evident with the 1938 hurricane that caused such devastation in New York and New England.

It took World War II to push the hurricane forecasting system another giant step. In 1943 the primary hurricane forecast office was moved from Jacksonville to Miami. There the Weather Bureau, the U.S. Navy, and the Army Air Corps combined to establish a joint hurricane warning service. After the war the Air Corps withdrew, leaving the Navy and the Weather Bureau to track and forecast hurricanes. The inclusion of the Navy allowed the Weather Bureau to use "hurricane hunter" aircraft to study tropical storm systems and to keep better tabs on the storm tracks. In 1965, after the Navy had moved its unit to Jacksonville, the Weather Bureau formulated a guide to further improve its hurricane forecasting ability.

During the late 1960s and in subsequent decades, the

Weather Bureau consolidated its hurricane warning system, streamlining its procedures and bringing everything under the umbrella of the National Hurricane Center in Miami. This allowed all hurricane track and intensity forecasts to be formulated and issued from one location. In addition, the NHC was given the responsibility of educating the public about hurricanes and their dangers.

THE STAKES RISE

The populations of the hurricane-prone areas of the United States increased dramatically following World War II. For example, the population of the coastal counties of Texas grew from about 1 million in 1940 to over 4 million by 1980. In Florida the population growth was even more extraordinary. The figures jumped from about 1 million residents in the southern coastal counties of the state in 1940 to over 9 million by 1980. In addition, the population of the eastern seaboard coastal areas, from North Carolina up to New England, increased also. Development in such areas made potential storm damage an economic as well as a human disaster. Also, the rapid growth of such areas brought in many new residents who had no experience with hurricanes, who were unaware of the storms' potential for destruction.

Paradoxically, the improvements in the hurricane warning system resulted in a dramatic decline in deaths between the years 1900 and 1980, yet the property damage soared during the same period (allowing for inflation). Between 1900 and 1910 over 8,000 people died from the devastating effects of hurricanes in the United States. By the decade 1980–1990 that number had dropped to 161. In contrast, property damage for the decade 1920–1930 totaled less than $2 billion, but rose to over $15 billion by the end of the 1980s.

In human terms, then, the growth of the hurricane warning system worked. It saved lives, and it saved property. The increase in property damage was a direct result of the growth of the population in the areas prone to hurricanes. But as recently as 1992 Hurricane Andrew swept ashore just south of Miami and caused death and destruction. It has become obvious that despite the strides made in the hurricane warning system over the years, there remains much to be done.

Accurate hurricane forecasts depend on an adequate source of data and its timely dissemination. The New England hurricane of 1938 would have caused much less destruction had the residents been properly forewarned. For many years the ability to forecast hurricanes was hampered by the lack of observations and the inability to communicate data between the field offices and forecast offices. Before 1935, when Congress appropriated funding to set up a special teletypewriter network, communications were made by the existing telegraph system. Congress had intended the new teletype system to link all of the Weather Bureau hurricane forecasting offices, which at that time were in New Orleans and Jacksonville, with smaller units in between.

Unfortunately, money ran out before a link could be made between Jacksonville and points farther north, including the main Weather Bureau office in Washington, D.C. Eventually, the linkup was made, and the teletype remained the main means of communications between Weather Bureau offices until the end of the 1970s. At that time the National Weather Service introduced the Automation of Field Operations and Services (AFOS) system as its main means of communications. The AFOS system permits computer displays of weather data. Its use resulted in the generation of the graphics you see on your nightly television weather reports.

At the same time the National Hurricane Center was improving its communication capability, it was also taking advantage of the technological advances made in the instruments used to monitor the atmosphere. In 1937 the weather service initiated a radiosonde network that enabled hurricane forecasters, for the first time, to analyze tropospheric currents that steer the hurricane.

By the end of the 1980s, the National Hurricane Center could utilize data received from satellites, from ships and aircraft, from ground stations throughout the world, and, with the aid of sophisticated computer systems, produce forecast models to aid them in predicting the severity and path of tropical storms. But there was another important source of data unique to hurricane forecasts: aircraft reconnaissance.

FLYING INTO THE EYE

No one is quite sure whose idea it was to fly into hurricanes for the purpose of gathering information. But historians do know that during World War II the U.S. Navy suffered massive losses from two Pacific typhoons. This precipitated the first recorded premeditated flight into the eye of a hurricane on July 27, 1943, by Air Force Col. Joseph P. Duckworth. Duckworth made two flights into a hurricane off the coast of Galveston, Texas, flying an AT-6 trainer aircraft. He was accompanied on one flight by an Air Force navigator and on the second flight by a weather officer.

Duckworth's flights began what became a long-running program of the U.S. Air Force: the tropical cyclone reconnaissance mission of Air Force weather reconnaissance units. This program was formally initiated in February 1944, with regular flights into the eyes of hurricanes conducted by Army Air Force and Navy aircraft, dubbed "hurricane hunters."

At first such flights were equipped with only primitive radar, but as technology improved, so did the instrumentation. With the introduction of Doppler radar and Inertial Navigation Systems (INS), hurricane hunting became a valuable tool for hurricane forecasters. By flying into hurricanes, scientists could gather information unavailable by other means: the moisture content of the clouds, the direction of the winds, the location of the precise center of the storm, and the direction and velocity of the surrounding winds. Such information has improved the understanding of the structure and characteristics of hurricanes.

But it isn't enough to be able to gather data while inside such storms. The data must be communicated to the storm center headquarters where it can be analyzed and the resulting forecasts disseminated to the public. In the early days of such flights, scientists on board the hurricane hunting aircraft reported back to the forecast center via radio. Recently, however, the aircraft and headquarters have been linked directly by a satellite and computer setup that allows them to communicate directly the digitized radar data obtained in the aircraft.

Of all the changes in hurricane forecasting that have occurred since midcentury, the most profound and far-reaching has been the development and operation of satellites. With the launching of the first geostationary satellites in the mid-1960s, and the development of the GOES series a decade later, hurricane forecasters now have a truly remarkable tool: a constant picture of the Atlantic Ocean from space. They can now see a tropical storm develop from infancy off the coast of Africa and follow it across the sea to the Caribbean. Although the data received from satellites is more qualitative than quantitative, such data have proved invaluable when used in association with radiosonde findings.

THE NATIONAL HURRICANE CENTER TODAY

The NHC today is composed of four units. The Hurricane Specialist/Forecast and the Tropical Satellite Analysis and Forecast units are concerned with the day-to-day analyzing and forecasting of tropical storms and hurricanes. The Techniques Development and Applications unit is similar to a Research and Development arm of any large corporation; it concerns itself with developing new techniques and applying them to daily operations. The Communications/ Charting and Computer Operations unit provides support for the two analysis and forecast units and acts as liaison to the public users of hurricane forecasts.

The NHC today provides what they call "product" to civilian and military and international users. Product means tropical and subtropical oceanographic forecasts of tropical storm development. Forecasts are issued routinely on a regular basis and, when tropical storms are developing, more often. The NHC's area of responsibility includes the tropical and subtropical portions of the Atlantic and Pacific Oceans from 32° N to the equator east of 140° W, an area that includes the Caribbean Sea, the Gulf of Mexico, and adjacent land areas. The NHC is also designated a Regional Specialized Meteorological Center (RSMC) under the auspices of the United Nations World Meteorological Organization.

Even before a hurricane appears on the horizon, the NHC has begun its forecasting process. The first step is to determine the vulnerability of coastal locations that could be in the storm's path. Using a model called SLOSH (Sea, Lake, and Overland Surge from Hurricanes), scientists initially use bathymetry data (i.e., the measurements of the depth and configuration of bodies of water) and topographical maps of the ocean floor to determine what kind of storm surge and flood damage would occur in a particular area if a storm were to hit in that location. Then they enter historical data on the results of past storms' impact on each area. By running various scenarios through the computer, scientists can get predictions for storm surges for each coastal area.

The NHC runs anywhere from 250 to 500 storm simulations for any given area. These are compiled and organized so that even before the storm hits shore, communities that may lie along its path receive these maps detailing just where rising water may do the most damage, assuming several possible storm tracks. Such information has proved invaluable to civil defense agencies in the affected communities who make evacuation plans and prepare for the storm's arrival. Ideally, these maps of possible storm surge damage arrive in time for the affected areas to be evacuated.

Hurricane forecasters use the same tools as other weather forecasters. They monitor changes in air pressure at different levels, wind direction and velocities, air temperatures at varying levels. Another important part of the hurricane forecast arsenal is the variety of on-site observing systems composed of buoys and ships at sea. Since the 1970s, the Navy and the National Weather Service have maintained a series of oceanic buoys along the Caribbean and Atlantic coasts. The buoys represent the only means of monitoring surface conditions on a daily basis. The advantage of buoys over ships for recording surface conditions is obvious: buoys remain in place no matter what the weather conditions, whereas most prudent ship captains try to maneuver out of the way of approaching storms. Fully automated weather buoys are able to supply continuous readings of weather conditions that are invaluable in forecasting approaching tropical storms.

The U.S. Navy first started using oceanic buoys in the late 1960s. The first system deployed by the National Data Buoy Center in support of hurricane forecasting was set up in 1972. According to Dr. Robert Sheets,[17] the Director of the National Hurricane Center, data provided by these systems have become a dependable and routine part of the daily analyses. They are also now a vital part of the hurricane warning system. To date, they are the only means of making nearly continuous direct measurements of surface conditions over these oceanic areas. Figure 18 shows observations from Hurricane Kate for the period when the hurricane passed over buoy 42003 in the eastern Gulf of Mexico. Of particular interest in these data is how the winds died down in the middle of the passage of the storm because the eye passed very close to the location of the buoy. The buoys measure both the average wind speed and the highest gust recorded at a particular time. Both parameters are important. The average wind speed is important for determining the actual dynamical structure of the hurricane and is most important for verifying how well the computer forecasts have performed. On the other hand, the strength of the gusts is most significant for determining how much damage can be done by the storm. Much greater automation of surface observations is planned for the future.

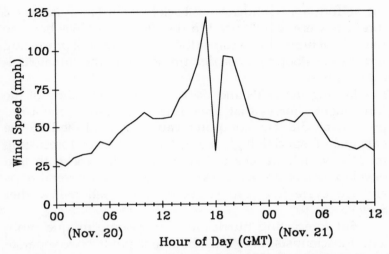

FIGURE 18. The wind speed of a hurricane measured by a buoy. The wind speed of Hurricane Kate was measured continuously by Buoy 42003 in the Gulf of Mexico between midnight GMT, November 20, and noon GMT, November 21, 1985. The passage of the hurricane's eye is easily seen at 18 GMT (noon, local time) on the 20th.

The data derived from satellites, radar, air reconnaissance, buoys, and ship reports are then run through a computer modeling program which predicts a storm's track and whether or not it will intensify or weaken. The models are similar to the ones described in Chapter 3 and are based on the same methodology first proposed by L. F. Richardson. One of the primary differences for hurricane modeling is the inclusion of ocean temperature, since much of the energy that drives a tropical storm is derived from the warm waters of the tropical oceans. Some statistical methods have also been developed based on a climatological knowledge of the performance of hurricanes back to 1886. These statistical forecasts were widely used in the 1950s and 1960s because the field of numerical weather forecasting was not producing forecasts as good as those determined from the climatological data base.

During the 1970s and 1980s, our understanding of the fundamental forces that drive hurricane development improved, as did our computer capability. This led to the creation of special high-resolution numerical forecasting models, which we use currently to predict the path and development of storms. These forecasts involve several steps and use several forecasting elements of the National Weather Service.

But forecasting hurricanes poses some unique problems for scientists. They must be able to predict two separate aspects of the storm: the environment surrounding the storm, which is most critical for determining the movement of the storm, and the actual structure of the storm itself. The National Hurricane Center uses data compiled by NMC models described in Chapter 3 to obtain information pertaining to the storm's environment, in addition to models obtained from the European Center for Medium-Range Weather Forecasting (ECMWF) and the United Kingdom Meteorological Office (UKMO).

Improvements to forecast models often take quantum steps. One such step especially pertinent to hurricane forecasting was increasing the horizontal and vertical resolution. Another was installing a new computer that can handle more detailed physics, such as how heat that is stored in the warm waters of the tropical oceans translates into more energy available for storm growth. Therefore, at certain times, the forecast products of the ECMWF or the UKMO may provide analyses that are more insightful than those put out by the NMC. That's why the NHC uses the output from the three best forecasting centers in the world before making its best determination as to what the surrounding environmental conditions will be at a later time.

The other aspect of hurricane forecasting involves detailed analyses of the tropical cyclone itself. These analyses involve all available satellite, reconnaissance aircraft, buoy, radar data, and ship observations to determine present and past motion, wind and pressure distributions, and other meteorological fields.

Typically, the NHC runs five to seven tropical cyclone track forecast models for each forecast cycle. In addition, there are two intensity forecast models used to predict the intensity of the storm. These intensity models are run for each case. These models are the basis for the NHC's 72-hour tropical storm forecast. Hurricane Jerry, which occurred in October 1989, provides an example of just how difficult it is to forecast a storm's progress.

Figure 19 shows the 72-hour model predictions for Hurricane Jerry for October 14, 1989. At that time the center of the storm was located in the Gulf of Mexico, near the coordinates 25° S, 93° W. The results for three different numerical forecasts are shown in this figure. All told, eight forecast models were run for Hurricane Jerry, but only three are shown here for the sake of clarity. The models are designated by various aspects describing some of the heritage behind their development. The model "EXPSAN83" (as well as two others) forecasted Jerry to make landfall near the Texas–Mexico border, whereas the model "CLIPPER" forecasted landfall in extreme eastern Louisiana. Another model, called "HURRAN" (not shown here), is based on statistics from previous hurricanes, and showed Jerry making landfall in western Florida.

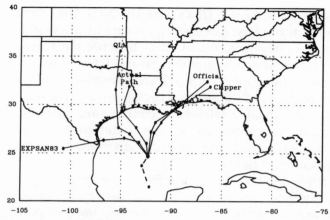

FIGURE 19. The forecasted trajectories of Hurricane Jerry. As many as eight different prediction models are used to forecast the path of an approaching hurricane. This particular figure shows the output from three of these models. After considerable analysis, a team of meteorologists at the National Hurricane Center comes up with an "official" 72-hour forecast which it relays to the public. The actual path of Hurricane Jerry is shown as the track labeled "Actual Path." This example illustrates the difficulty involved with forecasting the track of a hurricane, even with the availability of sophisticated computer models. Fortunately, the 24-hour forecast for Hurricane Jerry was considerably better and the city of Galveston was able to prepare for the arrival of this hurricane.

The staff of the NHC examined the output from all of the model calculations and then decided to issue an "official" forecast for the next three days that showed the hurricane making landfall in eastern Louisiana. In this particular case, the Hurricane Center's three-day forecast was not very good, as indicated by the track labeled "Actual Path" that Jerry took. As it turned out, the best model prediction for this particular case was the "QLM" model (and another model called "VICBAR" not shown on this figure).

On the other hand, even though this particular three-day forecast was not among the best ever produced by the NHC, Hurricane Jerry turned out to be a minimal hurricane with sustained winds of 75 mph and peak winds of 100 mph. Consequently, despite the fact that the Galveston area had only an 8-hour warning, it proved to be enough time to allow for adequate preparations, since the storm was relatively small.

The NHC can be proud of the outstanding job it did in forecasting the arrival of Hurricane Hugo, one of the most destructive hurricanes of the century. Total property loss was more than $10 billion with more than $7 billion in the continental United States and the remainder in the Caribbean. The total loss of life was 28 in the Caribbean and 21 in the continental United States, remarkably low considering the widespread destruction. The official 72-hour forecast for Hugo's arrival was within 154 nautical miles of its actual landfall.

The birth of Hugo was detected by satellite imagery on September 9, 1989, when a cluster of thunderstorms moved off the coast of Africa.[18] A tropical depression formed southeast of the Cape Verde Islands and moved westward across the tropical Atlantic Ocean at 20 mph, becoming a topical storm on the 11th and a hurricane on the 13th. On the 15th, reconnaissance aircraft examined Hugo east on the Leeward Islands (more than 2000 miles from the continental United States) and reported a wind speed of 190 mph at an altitude of 1500 feet. From this measurement, scientists estimated a surface wind speed of 160 mph. Hugo made a direct hit on the island of St. Croix on the 18th with a measured wind speed of 140 mph and grazed the eastern tip of Puerto Rico on the 18th with an estimated speed of 125 mph.

Leaving Puerto Rico, Hugo took aim at the South Carolina coast. Reconnaissance aircraft measured 160 mph winds at 12,000 feet and estimated surface winds of 140 mph. Downtown Charleston reported sustained surface winds of 88 mph with gusts to 108 mph. The strongest winds associated with Hugo were about 20 miles northeast of Charleston where sustained surface winds were 120 mph on the morning of September 22. Even in the town of Camden, more than 120 miles inland, it was estimated that one out of every four trees was blown down. The force of the storm snapped trees with two-foot diameter trunks 10 feet above the ground.

Once a storm looks as if it may land on the coastal United States, the NHC issues, along with the 72-hour forecast, the SLOSH maps showing predicted storm surges and water levels for each area. The storm surge along the South Carolina coast was 8–10 feet in the Charleston area and as much as 20 feet near the south end of Bulls Bay, about 25 miles up the coast northeast of Charleston. Without adequate warning, it is fair to say that tens of thousands of lives would have been lost because the population would not have been evacuated from these areas.

Until Hugo, the populace along the East Coast of the United States had become somewhat complacent despite numerous warnings by Bob Sheets, the current director of the Hurricane Center, and his predecessor, Neil Frank. They warned that it was only a matter of time before "the big one" would cause unprecedented damage to a coastal region that had grown from nearly nothing to a major metropolitan area over the past three decades. In south Florida, the population in the Miami–Ft. Lauderdale area had grown from less than 300,000 to nearly 3 million between the early 1950s and 1990. In the 1960s, Hurricanes Cleo and Betsy brought down trees, but between then and 1992, south Florida had been spared the devastation of a major hurricane. The good fortune of southern Florida ended in August 1992 with the arrival of Hurricane Andrew.

On August 14, satellites spotted a tropical disturbance in the Atlantic Ocean and the tropical storm was named "Andrew" the next day. This first storm of the season was nothing out of the ordinary. The NHC watched Andrew as it moved westward toward the United States. Although the tropical disturbance initially appeared to be quite strong, it was slow to develop and it wasn't until August 22 that the Hurricane Center upgraded the storm to hurricane status, meaning that its winds had reached the 74-mph plateau. By midnight on the 24th, with Andrew's eye still 110 miles from Miami, aircraft reports and satellite images showed the storm going through cycles of strength. The eye grew smaller and the winds stronger, then the eye enlarged and the winds weakened.

Over the previous several years, researchers had observed this phenomenon many times, and the forecasters at the NHC wondered which cycle the storm would be on when it crashed into Florida. At 1:30 A.M. on the 24th, radar showed that hurricane-force winds extended only 30 miles from the eye, implying that Andrew's strength was only about half of what was observed in Hugo in 1989. Andrew strengthened considerably in the next several hours. For the first time since it was built in 1979, the National Hurricane Center in Coral Gables was able to witness the onslaught of a major hurricane "from ground zero."[19] By 4:00 A.M., the dials on their instrument panels showed the rooftop anemometer spinning at 115 mph with an occasional stronger gust. By that time the building was shaking as if in the midst of an unending earthquake. Around 4:45 A.M., everyone heard a large "thump" as the radar antenna on the top of the roof crashed from its 10-foot tower. The anemometers showed a gust to 150 mph. At 5:50 A.M. the anemometer itself broke after withstanding a gust of 164 mph. Closer to the storm's center, gusts of 200 mph have been estimated. The accompanying storm surge was 17 feet. By the time it was over, Hurricane Andrew went down in history as the country's costliest natural disaster with a bill approaching $30 billion. The loss of life was kept to 65 people, including 43 in Florida, a remarkably low number considering the devastation that left 160,000 people homeless. The loss of life could well have been in the thousands, had not the National Hurricane Center been on top of the storm's development. The damage would also have been much worse had the center of the storm hit Miami head-on, about 40 miles north of the path that it did take.

THE FUTURE OF HURRICANE FORECASTING

Despite such examples, there is still room for improvement. The strength and intensity of Andrew highlight the continual need for basic research so that scientists can better understand the detailed and complex dynamics that go into the development of a hurricane. Why did Andrew grow from a relatively unremarkable hurricane to one of such dramatic proportions in barely more than two days? Scientists will use these observations of Andrew for years to come so that such dramatic development can be understood, and, we hope, even predict earlier the next time such an event will happen.

Even within the storm, some new theories have been developed that help to explain why there was such rampant destruction left in the wake of Andrew. One theory suggested by Professor Ted Fujita of the University of Chicago is that smaller vortices spin off the main winds as the hurricane moves.[20] If the wind speed of an individual vortex reaches a speed of 40 mph, the advancing edge of such a vortex embedded in an overall hurricane wind of 160 mph would account for an observed wind of 200 mph. Even structures designed to be "hurricane-proof" in winds of 160 mph would be toppled in the 200-mph winds. Fujita has analyzed damage from hundreds of tornadoes and is one of the leading experts in the field. Although these smaller scale vortices would not be considered tornadoes (on the other hand, tornadoes *are* often observed accompanying hurricanes), the pattern of destruction observed by Fujita would be consistent with the existence of such vortices.

Figure 20 shows how the NHC's forecasts for the loca-

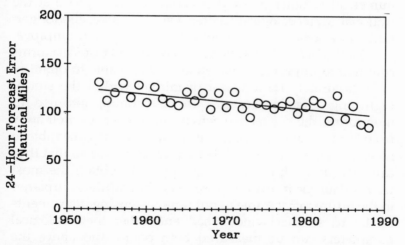

FIGURE 20. Twenty-four-hour forecast error of hurricane position over the past three decades. Plot shows yearly average error. The ability of the National Hurricane Center to forecast the position of a hurricane has steadily improved, a fact illustrated by this plot.

tion of hurricanes have improved since the 1950s. Whereas the average position of forecasted hurricanes (24-hour forecast) was about 130 nautical miles in error in 1954, the forecast error by the late 1980s was only about 100 nautical miles. Forecasters and numerical models continue to suffer from the lack of quantitative data over the tropics and subtropics. Therefore, analyses require manual interpretation of qualitative information. The GOES-NEXT generation of satellites is expected to provide more accurate and higher-resolution measurements than presently available from geosynchronous satellites. However, much of the information available to analysts will still be qualitative in nature. We can still expect that the best possible analysis for the tropical and subtropical regions will involve a bit of art. In other words, we need a human forecaster to modify and interpret even the best computer-generated initial analysis. As the field of numerical weather prediction continues to improve at the National Meteorological Center, we can expect the longer-range forecasts (36 to 72 hours) for the tropics to improve.

Methods for observations in and around tropical storms continue to improve. New sensors will be able to quantify more accurately the amount of rainfall within the storm. Lightning detection systems are also coming into use for monitoring the convective activity in hurricanes well away from land. New aircraft capabilities include airplanes able to receive satellite images while they are in the air so that they can plan and adjust their flight plans to obtain the most useful data for input into forecasts. In addition, airplanes in the future may be able to put remote-sensing instruments on board so that wind speed and other meteorological parameters can be measured both below and above the airplane's flight level.

Even though remote-sensing technology continues to

advance, the use of satellite-based sensors in the core of the hurricane is rather limited, partially because of the poor resolution provided from orbital altitudes of 400 miles or more. In Chapter 4, we discussed the new breed of laser sensors that will be able to probe the atmosphere looking down from high-flying fast airplanes that take wind and moisture measurements from an altitude high above the hurricane. Using these airborne instruments, meteorologists can obtain instantaneous height profiles of these meteorological variables. Thus, high-resolution two-dimensional (height versus horizontal distance) data fields can be obtained along the direction of flight of the airplane. By flying several crisscrossing patterns over the storm, scientists will be able to obtain impressive high-spatial-resolution three-dimensional depictions of the storm that have been heretofore unobtainable.

FORECASTING THE HURRICANE SEASON
BEFORE IT HAPPENS

One of the interesting features about hurricane activity in the Atlantic Ocean is that there is such a wide variability in the number of tropical disturbances that occur there from one year to the next, a situation scientists define as *interannual variability*. The study of this phenomenon has been the focus of research of William Gray, a professor of atmospheric science at Colorado State University for more than a decade. Every year, Professor Gray issues his hurricane forecast for the upcoming hurricane season (defined as June through October) on the first of June.[21]

What are the primary factors that determine how many hurricanes will form in the tropical Atlantic during the year?

Although quite a few factors are involved, Gray has found that the two most important ones are the strength of the winds in the lower stratosphere (between 65,000 and 75,000 feet) in the tropics, and the amount of rainfall in western Africa. Unlike the weather patterns at middle latitudes, which are dominated by seasonal variability (i.e., winters are cold and summers are hot), weather in the tropics is not strongly influenced by the seasons since low latitudes always receive a relatively large amount of solar radiation, regardless of the time of year.

On the other hand, weather patterns in the tropics are influenced by a phenomenon called the *quasi-biennial oscillation* (or QBO). What scientists have found is that certain meteorological patterns in the tropics occur at approximately two-year cycles. But it's not exactly two years: it's more on the order of 26 to 29 months, which is why scientists refer to this feature found in the tropics as the *quasi-biennial oscillation*. One of the meteorological parameters that seems to be most influenced by the QBO is the behavior of the winds in the lower stratosphere. What Gray has discovered is that when the winds in the lower stratosphere are in a westerly phase during the hurricane season, the storm formation potential in the tropics is much greater than when the winds are in an easterly phase. As it turns out, the QBO is quite an unusual atmospheric phenomenon because it can be extrapolated accurately ten months into the future. Research has shown that the westerly phase generally lasts 14–15 months whereas the easterly phase generally lasts 13–14 months. As a result, Gray can use this information to predict on June 1 how the winds in the stratosphere will be blowing and to determine what kind of storm activity can be expected between June and the end of October each year.

The behavior of the QBO during a particular year is only one of the pieces of information that Gray uses for his forecast. A second important piece of information is the presence and the strength of El Niño, a pool of warm water that often, but not always or regularly, is present off the west coast of Peru in the eastern Pacific Ocean. This pool of warm water has been associated with other weather patterns such as an extreme drought over the eastern United States in the summer of 1988, or the extremely dry conditions over India and tropical Southeast Asia in 1983. Both of these unusual weather patterns could be traced back to the deviations in the "normal" weather patterns that meteorologists found to be driven by the abnormally strong El Niños of 1987 and 1982, the two strongest of the 1980s. As it also turns out, the presence of a strong El Niño is generally associated with a lack of tropical storm development in the Atlantic Ocean. In his research Gray has gone back more than 40 years (to 1950) to correlate the presence of a strong El Niño with hurricane development in the Atlantic Ocean and his results show that these two years are the least active and the third least active hurricane-activity years of the ones he has analyzed.

Another factor that seems critical for determining the amount of storm activity in the tropical Atlantic is the amount of rainfall along the coast of western Africa that borders the Gulf of Guinea. In some of his latest research, published in September 1992 in the scientific journal *Weather and Forecasting*, Gray has determined that the amount of rainfall in western Africa may even be more crucial than the influence of El Niño.

These relationships uncovered by Gray's research again highlight the complex nature of the earth's weather machine. Why should winds at 75,000 feet, warm water in the Pacific, and rainfall in western Africa be tied to the formation of hurricanes that will have the most impact on the people living in the eastern and southern United States? The answers are just now coming to light. Perhaps one or two important factors in some other far-removed region of the world will be uncovered as better global measurements are made routinely from space. The potential is there. The possibility of predicting the likelihood of these killer storms months in advance is what makes the research to be conducted in the next few years so exhilarating and challenging.

SIX

Forecasting Severe Weather: Bouncing New Ideas Off the Subject

Perhaps the most awesome storms on earth are tornadoes. Who can forget in "The Wizard of Oz" the sight of the black funnel sweeping down out of the sky as Dorothy tries to get into the storm cellar. Sure it was magic, just a nylon stocking filled with sand and a bowl of dust, but that didn't detract from the illusion. Up there on the screen it looked like the real thing, giving the moviegoers from New York to Los Angeles a taste of the awesome power of a twister.

Tornadoes are the most powerful storms on earth. They are so powerful, and so fast moving, that accurate measurements of their awesome strength are yet to be made. The air pressure inside a tornado is impossible to measure precisely because by the time a barometer has registered it, the tornado has passed. But scientists have estimated that pressure within a tornado cell can be as low as 60% of the normal pressure, or down to less than 20 inches of mercury (most barometers don't show a scale below 28 inches of mercury).

FIGURE 21. Tornado observed in Cimarron County, Oklahoma, 17 miles north of Boise City. (Copyright © Gregory Stumpf, courtesy of Gregory Stumpf.)

Wind speeds in a tornado funnel are likewise impossible to record with any accuracy. In Topeka, Kansas, for example, the local weather station's anemometer had recorded a wind speed of 90 mph before being hit by a piece of wind-borne debris. In Springfield, Missouri, a tornado passed over a weather station and blew one of the cups off the three-cup anemometer. A recent tornado in Huntsville, Alabama, carried winds powerful enough to propel a large automobile and the woman driving it into an electric substation, through cinderblock walls and a steel safety fence. A man in Kansas found himself defying gravity by being deposited 80 feet in the air, atop high-voltage lines (knocked out by the storm).

Evidence gathered during the aftermath of tornadoes indicates wind speeds can reach anywhere from 240 mph to the speed of sound. Currently, meteorologists measure the strength of a tornado by using the Fujita Wind-Damage Scale, derived by T. Ted Fujita, a professor of meteorology at the University of Chicago, who has been inspecting the aftermath of tornado destruction since the 1940s.[22] The Fujita scale characterizes tornadoes by the severity of damage potential. The catagories are as follows:

- F0: Light Damage (winds 40–72 mph)—knocks over chimneys and billboards, breaks branches off trees.
- F1: Moderate Damage (73–112 mph)—peels surface off roofs, moves mobile homes, destroys attached garages.
- F2: Significant Damage (113–157 mph)—snaps or up-roots trees, tears off roofs, destroys mobile homes, pushes boxcars over.
- F3: Severe Damage (158–206 mph)—removes roofs and walls from well-constructed homes, overturns trains, lifts and tosses cars, uproots most trees in a forest.

- F4: Devastating Damage (207–260 mph)—levels well-constructed homes; generates large airborne missiles, including cars.
- F5: Incredible Damage (greater than 261 mph)—lifts strong frame houses off foundations, sweeps them away, and dashes them to pieces; debarks trees; badly damages steel-reinforced concrete structures.

More than 700 tornado sightings occur each year in the United States alone, most of them in the area known as Tornado Alley, stretching from Texas to South Dakota in the central United States. This area sees more tornadoes than anywhere else in the world. Americans who live outside this region think of tornadoes as a midwestern phenomenon, but in fact tornadoes have occurred in every state of the union on every day of the year except for January 16. Why that particular date has seen no tornadoes is one of the great quirks of weather science.

During modern times (since National Weather Service records have been kept) the year 1992 saw the most tornadoes ever—1,293. Of those, only one was classified as F5, 13 (1%) as F4, 43 (3%) as F3, and the remaining 96% as F0–F2 tornadoes. The worst single outbreak of twisters occurred on April 3–4, 1974, when 148 tornadoes swept through various parts of the country, killing 315 persons and injuring more than 5,000 others.

The area of the central United States known as Tornado Alley has some unique features that make it a breeding ground for tornadoes. The Rocky Mountains to the west produce an ample amount of subsiding air that flows over the mountains and onto the Great Plains. This creates upper-air inversions (i.e., the upper air is warmer than the surface) that affect cumulus cloud formation. The Bermuda high off the East Coast circulates warm, moist air from the Gulf of

FIGURE 22. The destruction from a powerful tornado demolished everything in this house except the interior shower. This struck Catoosa, Oklahoma, on April 25, 1993, and killed eight people. It was rated F4 on the Fujita scale. (Copyright © Gregory Stumpf, courtesy of Gregory Stumpf.)

Mexico northward into the Great Plains. The meeting of these two contrasting air masses provides the setting for the creation of severe thunderstorms and their attendant tornadoes.

Although tornadoes and hurricanes are both cyclonic storms with high winds and rain, they are actually quite dissimilar and occur under differing weather conditions. As we noted earlier, hurricanes form in the tropics along the tropical convergence zone, then travel across the sea for days or weeks while they grow into massive storms, sometimes encompassing hundreds of miles. Although this makes hurricanes larger and more potentially destructive, it also makes them easier to track and to prepare for.

Perhaps for the victims of strong winds and storms it makes little difference whether the culprit was a hurricane or a tornado. But the two types of storms are very different. Hurricanes cover a larger area and thus can do greater damage per storm, but their winds generally do not approach the velocity and destructive power that tornado winds do. So although tornadoes are smaller in scale, they are stronger in intensity.

The distinguishing characteristic of tornadoes is their suddenness: a twister can dip down from a thunderstorm, blaze a trail of destruction for a couple of miles, then retract back into the thundercloud in a matter of minutes. It is relatively easy to forecast the possibility of tornadoes, because tornadoes nearly always occur under very specific conditions. Conversely, there are a wide range of atmospheric conditions under which tornado formation *cannot* occur. But it has so far proved almost impossible to accurately predict where and when tornadoes will actually strike the ground. Recent advances in technology, however, may soon open up the possibility of predicting such occurrences.

TORNADO FORMATION

The exact causes of tornado formation remain a mystery. We do know what meteorological variables are necessary to create the storms, but those variables are not a guarantee that tornadoes will form. For a tornado to form, it needs a warm, moist, and unstable atmosphere, the kind that produces thunderstorms. As we noted earlier, an unstable atmosphere is one in which the vertical structure of temperature is such that strong upward vertical motions of the air can occur. Another factor necessary is wind shear at the upper levels. What this means is that the speed or direction (or both) of the wind in the upper atmosphere (typically 30,000–40,000 feet) are considerably different from the winds lower in the atmosphere. The jet stream, which is often very strong as it comes down around the Rockies into the Great Plains, helps create a strong cyclonic rotation on a very large scale that can feed into the winds within a tornado.

Research has shown that most tornadoes are associated with surface temperatures between 65 and 84°F, and with a dew point above 50°F. Dew points are a measure of humidity, the temperature at which water vapor will condense. The higher the dew point, the more humid the air. The dew point reading is a better indication than temperature alone of how the air "feels." The higher the dew point, the more uncomfortable it feels for human activity.

In general, tornado development is related to unseasonably warm, moist air and a contrasting air mass of colder and drier air. What gives a tornado its destructive strength is the whirling wind, the force behind the "twister." This intense twisting, circular motion of the wind can be better understood if we look at a basic law of physics known as *the law of conservation of angular momentum*.

This law governs the interaction of scales of motion in the atmosphere and elsewhere. It states that the momentum of any moving mass, which is simply its speed multiplied by its mass, must remain constant for that moving mass as it travels through the atmosphere. The momentum can only be increased by forces that are acting on it from outside of it. For example, a football player's momentum is determined by his speed and his mass. The same law states that movement in a circular path is directly related to the radius of curvature; i.e., in order to maintain constant momentum, any change in the radius must result in a corresponding change in velocity. You can see this clearly when a figure skater does a spin and the rate of spin increases when the skater draws her arms in toward her body. Or, when you twirl a ball attached to a string around your finger, as the string wraps around your finger, thus decreasing the radius, the velocity of the ball increases.

Tornadoes do not form in a vacuum. Their development is linked to the development of thunderstorms—when conditions are ripe for the manifestation of the rotating winds inside such thunderstorms, tornadoes may occur. The National Weather Service categorizes thunderstorms as severe when they exhibit extremely high winds, hail, and the possibility of tornadoes. Hail is a good indicator of a storm's strength, since it is formed when raindrops are hurled upward thousands of feet within a thundercloud, over and over, until the raindrops freeze and fall to the ground as hailstones.

Most thunderstorms, however, are relatively benign. They are nature's way of cooling off the atmosphere during an extreme heat buildup. Most such thunderstorms are fairly local in nature, their effect being limited to a small area, what weather forecasters call "scattered thunderstorms." Sometimes thunderstorms grow to quite a large size, covering miles across and thousands of feet high. It is the size of a thunderstorm that usually determines its severity. The larger the thunderstorm, the more severe the conditions are likely to be inside it.

For a thunderstorm to form it needs a localized unstable air mass. Thunderstorms can develop either independently, as part of a squall line associated with a front between two contrasting air masses, or as a multicell structure. A thunderstorm forms in the following way. First the localized air becomes unstable. This can occur because of heating from intense sunlight, such as your typical summer afternoon shower; or from contrasting air masses, which can occur when a cold front pushes into a warm, moist air mass. The basic process begins when a bubble of air, a "cell," becomes more unstable than the surrounding air and rises to form a cloud.

This unstable air is the most important factor in the creation of thunderstorms and tornadoes. If the air is stable there are few vertical air currents; if unstable there is a lot of mixing vertically and it is this mixing that creates severe thunderstorms. But what does it mean when the television weather forecaster talks about stable air? Stable air is what you see when you look up on those cloudless, clear, beautiful days—days of blue sky and high pressure. Whether in winter or summer, the stable air mass is characterized by how the temperature decreases as it rises in altitude. The energy that a parcel of air at a particular altitude in the atmosphere contains is defined by a property called its *potential temperature*. We know that as we go higher in the atmosphere, the temperature decreases. The potential temperature can be quantified by bringing that parcel back down to the surface and seeing what temperature it would have at the surface.

Furthermore, the energy that a particular parcel contains can be classified into two parts: its *kinetic* energy and its *potential* energy. The kinetic energy of an air parcel is readily determined by its temperature. Its potential energy is simply defined by how much energy it would possess if it were lowered to sea level. The *lapse rate* is defined as how temperature decreases with altitude. In an atmosphere that has the same potential temperature at every altitude, the air would be cooler by 6.5°F for every 1,000 foot rise in elevation. Thus, the potential temperature of a parcel of air with a temperature of 70°F at 1,000 feet would have a potential temperature of 76.5°F. This is analogous to how much energy any object would possess if it were dropped from a higher altitude—the higher above the surface, the more potential energy it contains. Thus, since a parcel that is lifted from the surface to a higher altitude must increase its potential energy, and conserve its total energy content, it must therefore decrease its kinetic energy, which is why air is colder the higher up in the atmosphere it is.

Another factor necessary for the development of thunderstorms is a high *dew point* temperature. Moist air is generally more unstable than dry air since more energy had to be expended to put that moisture into the air in the first place. One reason most tornadoes occur in the Great Plains is because this area of the United States is often under the influence of moist air from the Gulf of Mexico, which invades the plains as a moist "tongue" of air carried into the interior by the prevailing winds of a low-pressure area settling down in the Southwest.

A third factor is wind shear at upper levels. Many

thunderstorms move along below the jet stream, so that the circulation patterns in the upper levels support and enhance the circulation of the thunderstorm. In the tornado-prone areas there is often a low-level jet stream that flows north-ward above the moist tongue from the Gulf. When this jet stream meets the major east–west jet stream above it, a wind shear is established that creates a vortex around which severe weather, such as tornadoes, can develop.

Still another ingredient necessary for the development of severe thunderstorms is an upper-air inversion. Inversions, remember, are areas in the atmosphere where the temperature *increases* with altitude rather than decreases, as is normally the case from the surface upward to about 7 miles. Research indicates that prior to the development of severe thunderstorms and tornadoes, an inversion layer exists at about the 5,000-foot level for a short distance, and then the temperature begins decreasing again.

What this inversion layer does is suppress cloud development until a thermal (in the form of a small cloud) can break through the layer. Such a delay allows the resulting thunderstorm to grow much larger than it would otherwise, without the inversion layer. In much the same way that damming up a stream allows the water to build up pressure so that when it finally does break through the dam its force is greater than it would have normally been, the inversion layer allows the development of far greater forces by allowing the sun to heat the ground and lower atmosphere until it is hotter than the inversion layer. Such a cloud, when it does break through, can build up rapidly because the air in the cloud is already warmer than the air above. Because the temperature of the air that had just broken through the inversion is higher than the air in the environment in which it is now located, this new air parcel is buoyant and keeps on rising. The presence of an inversion layer often is the difference between the development of an ordinary thunderstorm and a severe one.

A tornado forms inside a thunderstorm cloud. As a rotary motion is set up by converging winds, the winds inside the whirlwind intensify as the funnel increasingly tightens into a smaller circle as the pressure drops because of the updraft created inside the whirling winds. As the radius of the twister decreases, the winds inside it further intensify to incredible speeds. It is these winds that cause the most destruction in a tornado. They are the most powerful winds on earth.

But winds are not the only destructive force in tornadoes. The rotary winds set up a cycle by causing intense updrafts, which in turn reduce the air pressure inside the twister. Scientists estimate that air pressure inside a funnel can cause a drop of atmospheric pressure by 8% in a matter of seconds. As an example of how destructive this can be, assume the pressure inside a house is normal at sea level, about 15 pounds per square inch. If a tornado sweeps over the house, with a pressure drop of 8%, that would mean that the pressure outside the house is only 13.8 pounds per square inch. The drop in pressure is so sudden that the pressure inside the house can't catch up with the pressure outside, especially if the doors and windows are closed. With a force equivalent to 60 or so tons exerted on the roof, many buildings blow their tops off during tornadoes as if several pounds of dynamite had been set off inside them. Wise residents of tornado-prone areas know to leave a window open during severe storms to prevent their roof from blowing off or their walls from blowing out.

Many tornado stories attest to the storm's destructive and awesome force. Between the loss of pressure and the extremely high winds, even a twister that is only a few hundred yards wide can do tremendous damage. Yet at the same time, some storms seem to have a gentle side, too. One twister touched down in New England and promptly destroyed a knitting mill. Thirty-five miles away, the storm deposited packages of knitting material completely intact and unhurt. One woman was sucked out an open window and deposited, unhurt, 60 feet away. In her hand was a phonograph record she was about to play. The title? "Stormy Weather."

CONDITIONS FAVORING THUNDERSTORMS

Thunderstorms, and their rain and tornado offspring, develop under several kinds of atmospheric conditions. In tropical and semitropical regions, localized thunderstorm activity results from the heating of land beneath an unstable air mass. Here the thermal updrafts caused by the sun's heating of the land build in intensity, creating thunder-clouds and a resulting rain shower, usually between noon and midafternoon. In some areas you can set your watch by the daily thundershower. Such showers, while containing lightning and heavy rain, are seldom of the severe variety and in fact serve to cool the surface atmosphere. The missing ingredient in the tropics is the lack of a jet stream, which normally is present only at middle and high latitudes.

On the other hand, surface heating is not generally present at high latitudes. Therefore, middle latitudes have both enough surface heating in the late spring and early summer, and the likely presence of a rather strong jet stream at this time of the year to produce ideal conditions necessary for the formation of severe weather.

In the middle latitudes, more severe thunderstorms seem to occur in the warm air sector of large-scale storms, or cyclones. The warm air sector is that quadrant usually to the southeast of the cyclone's center. Typical midlatitude cyclones feature a warm front trailing before it to the east, followed by a cold front to its west. In between the two fronts is the warm air sector. If the surface low-pressure cell lies below the jet stream trough, as is often the case in the Great Plains area known as Tornado Alley in the United States, then the winds aloft from behind the cold sector can outrun and override the warm air sector, creating ideal conditions for the formation of severe thunderstorms.

Another feature associated with many of the more severe thunderstorms is the squall line. Although individual thunderstorms can become severe, a squall line intensifies thunderstorm growth. Squall lines are associated with the midlatitude cyclone, where thunderstorms form in a line equivalent to where the cold air invades the warm sector south and east of the low-pressure center. Squall lines can contain several severe thunderstorms, and the relative proximity of these storms to each other seems to provide the energy to form new storms. The progression of the area of thunderstorm development generally moves in conjunction with the larger-scale frontal system as it is carried along by the upper-level winds associated with the jet stream. When squall lines exist in areas where other necessary atmospheric ingredients for severe weather are present, the likelihood of tornadoes becomes greater. Squall lines and resulting tornadoes can also result from hurricanes that have come ashore. As the larger storm's energy is dissipated over land, smaller cells spin off from the larger storm and can develop tornadoes.

FORECASTING TORNADOES

Since most tornadoes occur in the area of the Great Plains, that is where the most research on tornado forecasting has taken place. Tornadoes are the children of thunderstorms. Severe thunderstorms, even without a tornado, are capable of causing damage to crops and buildings, and creating flood conditions.

Thunderstorms originate in those peaceful-looking cumulus clouds that dot a postcard-perfect day in summer. Those small cumulus clouds that resemble cotton puffs in the sky are formed by warm thermal updrafts from the surface. For a thunderstorm to develop there needs to be warm, moist air, sunshine to heat the ground and set up strong updrafts, and unstable air at higher levels. As the moisture-laden air is carried aloft, it cools and the moisture condenses, creating a cloud. If the air is unstable enough, and if there is enough moisture, the cumulus cloud builds vertically, setting up its own structure of updrafts and circulation patterns that help the cloud grow.

In most cases, such a scenario will end with a thundershower, and the entire process can take as little as one hour. This is the type of scenario experienced in the tropics, where in some places you can set your watch by the daily afternoon thunderstorm. But in more temperate climates, such thunderstorms can grow to even greater heights and become severe. The severity of a thunderstorm generally depends on how high it can grow vertically before the upper-level winds create shear along the tops. Such storms thus create an internal structure that allows them to live a longer life than an ordinary thunderstorm and to develop higher winds, hail, and even tornadoes.

Pinpointing the exact location of where a severe thunderstorm and resulting tornado will occur is still not possible, but weather forecasters are able to predict a general area where the atmospheric conditions are conducive to severe storm development. To do this they need data from the upper atmosphere. They need to know if there is an inversion layer up there, and from what direction the winds are coming so they can determine if there is any wind shear. Isolated storms, i.e., those not part of a moving front or squall line, often develop on days when the mornings show clear skies and not a hint of a storm.

When the ingredients for a severe storm are present, forecasters at the National Weather Service issue a *tornado watch*. This simply means that the *possibility* of severe storms and tornadoes exists within the area. The area included in a tornado watch can cover thousands of square miles and several states. A *tornado warning* is issued for a more specific area. Tornado warnings are issued only after a tornado has been sighted. The area included in a tornado warning is that which lies within the projected path of such storms.

The center for forecasting severe storms and tornadoes is the National Severe Storms Forecast Center in Kansas City, Missouri. It is here where data are compiled and distributed throughout the 48 conterminous states. The NSSFC issues tornado watches and warnings and keeps records of the number of tornadoes spotted during the year.

Until the establishment of a central tornado and severe storm forecasting center, local weather stations and communities were left to their own devices to forecast such storms. The result was a yearly carnage of human life and property as tornadoes pounced without warning on local communities.

THE HISTORY OF SEVERE STORM FORECASTING

As more and more of the Great Plains became settled during the 19th century, the importance of severe storm forecasting rose. Settlement meant ranches and farms, houses and towns, people, livestock, and crops. One hailstorm is enough to wipe out a farm's entire crop of corn. One tornado can destroy every building in a small town, or it can skip and weave, totally obliterating one house while leaving the next one untouched. A tornado landing amid a herd of cattle can hurtle the beasts hundreds of yards and destroy the dreams of its owner.

During the latter half of the century, scientists were beginning to learn more and more about the dynamics of thunderstorms and severe weather. Despite this growth in knowledge, there was little progress in the ability to forecast such storms. The United States Land Office made the first attempt to set up a minor network of meteorological observation stations in 1817. Two years later the surgeon general of the U.S. Army took over and expanded the network.

But it wasn't until the availability of a new invention, the telegraph, in 1845, that weather stations worldwide could communicate with each other quickly. In 1847 the Smithsonian Institute inaugurated a voluntary network of meteorological observation stations for the purpose of "solving the problems of American storms." In the first year of its existence, the network included 150 stations, which grew to more than 500 by the start of the Civil War in 1860.

Unfortunately, the war resulted in a decline in such stations, and the network never regained its numbers after hostilities ceased. In 1873 the Smithsonian petitioned the U.S. Army Signal Corps to take over the weather observation network. In January of the following year the Signal Service began providing real-time weather observations used in the preparation of charts for the forecaster, called the "indications officer."

When General William Hazen took over the Signal Service in 1880, he hired seven scientists and established a research unit that he called the study room. The first task of the unit was the tornado studies—to collect and investigate tornado reports in order to develop procedures to forecast such storms. The man in charge of the project was Sergeant John P. Finley. He established a network of "tornado reporters" east of the Rocky Mountains, starting out with 120 reporters. By 1887 the number had reached 2,403. These were all volunteers, working as a separate group apart from the Signal Corps.

Finley began issuing tornado predictions in 1884, developing a chart that used the surface parameters instrumental in the formation of tornadoes. The chart was quite advanced for its time, employing such factors as temperature, dew point depression, wind direction and speed, and current weather. Unfortunately, just about the time Finley began formulating his predictions the Signal Corps issued a ban on the use of the word tornado in its forecasts, fearing that use of the word might cause undue alarm among the populace.

In the summer of 1884 the Signal Corps embarked on an ambitious project to make a systematic study of thunderstorms during the summer season. The project required more observing centers at closer proximity, stretching from the Atlantic Ocean to 102° W longitude (the Great Plains), and from 34° N latitude (the Carolinas) to the Canadian border. Within this area, post offices situated about 40 miles apart were enlisted to take thunderstorm observations for the summer. The Signal Corps distributed guidelines. The results were modest, but it was a beginning on the road to centralize data from the field to determine the best method of forecasting severe storms.

Meanwhile, in New England, the newly formed New England Meteorological Society initiated the same kind of study of thunderstorms for the summer of 1885, using 300 volunteer observers. The results of this study coincided with those of the larger study done by the Signal Corps: thunderstorms move parallel to the upper-level winds.

By 1890, the U.S. Congress passed a bill transferring the weather service to the Department of Agriculture. One of the first projects approved by the new civilian weather service was the investigation of thunderstorms during the summer of 1892. The investigation had two objectives: to improve the forecasts of such storms and to obtain better knowledge of their characteristics. The area selected for the study did not include the Great Plains, but stretched from the upper Mississippi valley eastward to the Atlantic Ocean.

At that time there were no guidelines enabling forecasters to predict thunderstorms from the daily weather charts. After the study was over, scientists learned that 90% of the thunderstorms observed occurred in a belt covered by the 30.00 inch-or-less isobar and at or near the 70°F isotherm. Scientists realized that there were two classes of thunderstorms: one class moved in a regular progression from west to east and was associated with a frontal system; another class, referred to as "heat thunderstorms," occurred over a large area without a definite path of progression. The latter were generally not as intense as those associated with a moving front.

The findings from that summer's investigation also produced the first substantial formula for predicting severe thunderstorms. In the report it was noted that the presence of a low-pressure area to the west, moving behind a high-pressure area, should be "watched with great care as the thunderstorm conditions are very liable to develop during the afternoon or evening." The report went on to state that "the sharp curvature of the isobars, especially where it touches or crosses similar sharp curves of temperature, has been found to be of value in forecasting thunderstorms." Today meteorologists call the sharp curvature of the surface isobars a "V-shaped trough," and it is still one of the good predictors of thunderstorm activity.

Using this new information, the weather service began forecasting thunderstorms. Verification in Ohio, Michigan, and New England showed an accuracy rate of between 69% and 86%.

Unfortunately, despite the success of the summer thunderstorm investigation, further research was abandoned between the years 1897 and 1916. Changes in the Weather Bureau administration and politics were major factors leading to the abandonment of research during this time. This lack of interest in tornadoes in the area of the world most productive of tornadoes led European scientists to wonder publicly why an area more afflicted with tornadoes than any other area of equal size on the globe should produce almost no literature on severe storms. Even more embarrassing, the Weather Bureau was one of the last in the industrialized world to accept the new theories of the Norwegian School, including frontal analysis and air-mass analysis. It wasn't until the late 1930s, in fact, that fronts began appearing on the Daily Weather Map released to the public by the Weather Bureau.

One reason for the lack of research and literature on tornadoes was the Weather Bureau's ban on the use of the word tornado, not officially lifted until 1938! Despite the ban, however, research on severe storms and tornadoes was being undertaken by a scattered group of meteorologists, and their findings were published in minor publications, for the most part ignored by the bureaucratically entrenched Weather Bureau.

It took a major war to awaken the Weather Bureau out of its somnambulism. As the nation prepared for World War II, the Weather Bureau was transferred from the Department of Agriculture to the Department of Commerce in 1940. The reason given was that the developing aviation industry was under Commerce Department jurisdiction and such an industry needed close links to weather forecasting capabilities. But the real reason was that by 1940 many factories were manufacturing explosives, and the management of such plants needed information regarding impending severe thunderstorms. Our war effort had already begun, even though it wasn't until the end of 1941 that the United States officially entered the war.

The Defense Meteorological Committee was formed to ensure that the Weather Bureau and the War Department maintained open channels of communication. The committee beefed up the research and development along the Pacific coast and in Alaska, and improved the weather forecasting over the Atlantic. It was this committee that perceived the importance of severe storm forecasts in and around areas where munitions and other defense plants were located.

The Weather Bureau organized a severe storm warning service to issue advisories to plant officials of impending thunderstorms. Their big fear was that lightning could set off gunpowder and other high-explosive materials used in the plants. What officials were most afraid of happening in such plants was fire. And since severe storms are usually accompanied by lightning, their fears were justified. The severe storm warning plan consisted of recruited observers stationed at several points within a 35-mile radius of the munitions plants. When these observers saw approaching thunderstorms, they would telephone either a central Weather Bureau station or directly to the factory in the path of the storm. By December 1942 100 networks had been established around plants manufacturing military supplies.

During the war strict censorship was imposed on the dissemination of weather information. But the military authorities did allow the broadcast of severe weather conditions. As the war effort increased and military bases and defense plants were opened across the country, the demand for information on severe storms increased. This led to the formation of experimental tornado warning systems in three locations: Wichita, Kansas; Kansas City, Missouri; and St. Louis, Missouri.

By war's end in 1945 the Weather Bureau had set up more than 200 severe storm observation networks, in addition to those set up specifically around military bases and defense plants. During this time the Weather Bureau issued a bulletin defining what was expected of the local forecaster. In essence, this bulletin gave the local forecaster guidelines on how to determine if the weather conditions warranted a warning or an alert, in much the same way that today differences exist between tornado watches and tornado warnings.

The feeling then was that severe storm alerts were more trouble than they were worth, since such an alert caused public alarm and panic, followed by indifference if a tornado never materialized. Yet tornadoes and severe storms were capable of damaging the war effort by wreaking havoc on air bases and munitions plants. During 1942, tornadoes caused more damage and loss of life than in any previous year. This prompted a Weather Bureau research project on the formation of tornadoes that evolved into more exact guidelines for weather forecasters to predict tornado activity.

In 1948 a tornado struck Tinker Air Force Base near Oklahoma City. The twister turned aircraft into twisted sculptures of metal. It leveled administration buildings and tore the roofs off of hangars. Altogether, 32 aircraft were destroyed and a number of buildings on the base had to be rebuilt. Tinker AFB was the headquarters for the Air Weather Service (AWS), the military offshoot of the Weather Bureau. The base commander called in two of the AWS's top forecasters and asked if it wasn't possible to provide better advanced warning of severe weather. The two forecasters, Major Ernest J. Fawbush and Captain Robert C. Miller, had already been working on a new, empirical technique for predicting tornadoes. They went to work and five days after the first tornado hit they predicted another one and were able to institute a warning that saved many of the planes the first tornado had missed.

Fawbush and Miller continued to work on their technique and to issue forecasts for the central Oklahoma area that were successful. Although their forecasts were issued only to Air Force weather offices, word got out that the Air Force was issuing tornado forecasts for their own troops but the Weather Bureau wouldn't do the same for the public. In 1950 the Weather Bureau ended up inviting Fawbush and Miller to its Kansas City office to discuss their forecasting procedures. After a couple of visits to the district office, and subsequent visits of Weather Bureau officials from Washington to Tinker AFB to discuss the Fawbush/Miller technique, the Weather Bureau decided that their procedures did not warrant any change in Weather Bureau policy.

Interestingly, the Weather Bureau's primary problem with the Fawbush/Miller tornado forecasts was that the latter failed to pinpoint tornado occurrence. In other words, what Fawbush and Miller were able to do was to predict that, under particular types of weather conditions, they could say that tornadoes were *likely* to occur. This inability to forecast the precise location and path of tornadoes continues to this day. The primary reason that only relatively large areas can be pinpointed by Fawbush/Miller and subsequent techniques is that these forecasts rely heavily on the structure of the atmosphere well above the surface, and the resolution of these measurements is limited to how far apart the upper-air stations are located. These stations are expensive to operate, and typically placed several hundred miles apart, which limits the resolution of the analyses derived from them.

In the early 1950s, urged on by a public clamor for better tornado forecasts, the Weather Bureau in Washington initiated what it called the Tornado Project. It was a program to study a theory of Dr. Morris Tepper that hypothesized tornadoes were more likely to occur along squall lines where cold fronts met warm air. He called this area a "pressure jump line" and theorized that where two of these pressure jump lines met was a likely area for the production of tornadoes. The Tornado Project was initiated in the Kansas and Oklahoma areas, using 134 observation stations. High-speed microbarographs and other state-of-the-art equipment were installed at each of the observing stations.

The end result after the first year was that there was "some evidence of compatibility between the observational data and the pressure jump line theory." The Tornado Project continued for several more years, its name changing and its area expanding with each new severe storm season. But the project proved to be a boon to the Weather Bureau, for during the time the project was in place much was learned about the formation and characteristics of severe storms and tornadoes.

Yet, during the early 1950s, the Weather Bureau still refused to issue tornado warnings to the general public. Finally, bowing to pressure from the press and public, the Weather Bureau in 1952 decided to begin issuing tornado forecasts for the public, despite its reservations about the efficacy of such forecasts. In the spring of that year, a conference was held in Washington attended by Weather Bureau chiefs from the entire area east of the Rocky Mountains. At the conference the participants devised an operational plan that included both forecast and warning procedures.

To streamline the dissemination of such forecasts, the Severe Weather Unit of the Weather Bureau became the Severe Local Storms Center (SELS). During the next several years the Weather Bureau employed several methods attempting to improve their severe storm forecasting capabilities, testing their research with the only laboratory they knew: the atmosphere over the severe storm-prone area of the Midwest. This was one reason the SELS unit was moved to Kansas City in 1954, where its meteorologists and researchers continued devising methods that would effectively predict severe storm and tornado activity.

In 1955 the Weather Bureau signed a contract with an experienced cloud-seeding pilot. The pilot, James Cook, owned an F-51 aircraft. The Weather Bureau equipped the plane with special instruments and hired Cook to pilot the craft as an observational platform to further study tornado formation. His job was to make observations of "temperature and humidity gradients in both the horizontal and vertical plane," as well as other meteorological elements as dictated by the Weather Bureau.

The project, named the Tornado Research Airplane

Project (TRAP), proved less than a success. The instruments aboard the plane needed constant calibration, and because of poor maintenance the plane missed many of its flights. But it did operate for about 30 days and flew in or around 36 tornadoes and severe storms. SELS researchers spent the following year analyzing the data collected from TRAP. The program continued for several years, eventually becoming part of a program initiated in 1960 that saw the Weather Bureau collaborate with the Air Force, Navy, NASA, and FAA. This program, called the National Severe Local Storms Research Project (NSLSRP), included the use of multiple plane missions into thunderstorms and squall lines.

With 13 aircraft, the new project became one of the largest field research programs ever conceived for the study of a geophysical phenomenon. The SELS program, based in Kansas City, became the basis for the consolidation of all severe storm and tornado research and analysis that eventually became the National Severe Storms Forecast Center.

FORECASTING TORNADOES: HOW NEXRAD WILL REVOLUTIONIZE OUR ABILITY TO OBSERVE THE RIGHT SCALE

During the growth of the National Severe Storms Center, and especially since World War II, much research was undertaken to improve the forecasting of severe storms and tornadoes. As was the case with the development of the National Hurricane Center, the Severe Storms Center became a laboratory and operating forecasting center at the same time, researching new equipment and methods while maintaining its function within the National Weather Service.

In the early days of the center's operation, they had established a primitive radar network through Texas, Oklahoma, Arkansas, and Louisiana using surplus radar from World War II. The radar worked fine, considering the alternative, but the problem was there was no system to transmit the radar's findings to other stations and to the national office in Washington. Radar reports were transmitted over a teletypewriter system on a "time available" basis, meaning an operator with an "active" radar report in his hands sometimes had to wait for an hour before the teletypewriter line was free for him to transmit his warning.

This situation was corrected in 1955 with the implementation of a new system dedicated strictly to radar reports and warning coordination. This led to the creation of the Radar Analysis and Development Unit (RADU) within the bureau in 1956. The RADU collected, analyzed, and transmitted hourly summaries of radar reports over the radar warning system. The control center for this service was merged with SELS in Kansas City.

A new generation of radar (the WSR-57, where the 57 denotes the year in which the system was developed) was installed by the National Weather Service during the 1960s and updated over the next two decades. Military radars (the AN/FPS-77) are somewhat more modern, but for the most part, digital technology that most of us have in our TVs and stereo systems is not contained within the design of most of the radars that are our first line of information for trying to predict where killer tornadoes are going to strike. Radar had proved an important tool in the forecasting of severe storms, but it had its limits when it came to predicting tornadoes.

One significant aspect of forecasting weather is that of scale. The science of modern meteorology actually began at the turn of this century in Norway where the Bergen School introduced the concept of the polar front. This concept established the importance of fronts and how the evolution of cyclones influenced the weather throughout northern middle latitudes, something we take for granted now. These cyclones and accompanying fronts carried with them characteristic weather as they moved from west to east: rain ahead of the fronts and generally clear and cold weather behind them.

Along with the polar front theory, forecasters began using synoptic weather maps. Synoptic maps were made from observations taken at the same time at all locations, another aspect of modern weather forecasting taken for granted now. With such observations, forecasters could study maps that depicted weather conditions on a far larger scale, one that encompassed thousands of miles and included all of Europe and the United States. Meteorologists call the processes that are resolved on this scale the *synoptic scale*. Observing networks in the United States, Europe, and other regions evolved to provide information so that accurate synoptic-scale forecasts could be achieved.

Severe weather and tornadoes, however, occur on much smaller scales. This is why the Weather Bureau, back in the 1950s, set up special observing networks in Oklahoma and Kansas, to allow for observations on a smaller scale. In fact, in 1957 Ted Fujita, then at the University of Chicago, wrote a paper describing a weather analysis that resolved scales on the order of tens of miles, rather than the larger synoptic scale of hundreds of miles. The title of his paper was "Mesoanalysis."

Fujita worked closely with Morris Tepper in the early 1950s to try to obtain an understanding of how the pressure jump feature, identified earlier by Tepper, evolved. To do this they needed to understand the atmosphere on a much finer scale than could be achieved through existing observational networks. What Fujita and Tepper's analyses described were features they called *mesohighs* and *mesolows*, which were analogous to the highs and lows seen on conventional weather maps associated with the generalized polar front theory. The "meso" features, however, were associated with the structure that evolved around squall lines.

Research on *mesoscale* meteorology took off during the 1950s and 1960s, since it was believed that a good depiction of the physical processes occurring on this scale would eventually lead to a reliable method of forecasting the precise location of violent thunderstorms and even tornadoes. What happens within thunderstorms and tornadoes occurs on an even smaller scale than the mesoscale researchers were prepared for. The physics and circulation processes of thunderstorms and tornadoes occur on a *microscale*, which meteorologists generally define as processes that take place on spatial scales of no more than a mile or so.

Research meteorologists view as most important not so much the individual scales as the *interaction* of the scales with each other. Such a study would provide an important key to the advancement of the science of meteorology. What the research meteorologist hopes, what he or she is always looking for, is the one key parameter, the diamond-encrusted nugget of information that could be used to predict the exact time and precise location for the formation of a tornado. One recent technological advance brings us one step closer to this hope: Doppler radar.

Even as the National Weather Service's first WSR-57 radar systems went into operation back in the late 1950s, an alternative radar system was already being conceived that would have the capability to "see" motions within clouds. The WSR-57 is very good at what it does. It can detect and track fast-moving, separated, "hard" targets like airplanes, which was its original purpose. It is also good at detecting pieces of ice and very large water droplets within clouds, which is the reason it had become so useful as a meteorological tool. The conventional radar systems, however, are not very successful at detecting "soft" targets, such as moisture-laden particles suspended in the air, which can number in the billions and even quadrillions. And they are not very successful at detecting targets which, thanks to the turbulent winds that whip the particles about, have motions within motions, the kind of winds found in severe storms. These situations are wholly beyond the capability of conventional radars to track.

During the past decade, however, new developments in radar technology have been successfully implemented. These new systems not only "see" soft targets, but also take advantage of the so-called "Doppler effect," a frequency shift first described by the 19th century Austrian physicist, Christian Doppler. Doppler was intrigued by his observation that the pitch of a train whistle changed as it approached an observer, being higher as it approached and lower as it passed. So, in addition to seeing these "soft" particles, the new radars being developed in the early 1960s were specially engineered to measure the Doppler frequency shift as the radar beam bounced off these particles. The new radar system, which eventually became known as *NEXRAD*, was capable of measuring the velocity of anything moving toward or away from it even when the movement was quite slow, thereby achieving a new level of information that could not be provided by conventional radar.

The new Doppler radar system was tested at the National Severe Storms Laboratory in 1971 and provided exciting results immediately. Not only did NSSL meteorologists have little trouble identifying mesocyclones with it, but the digital data generated in the course of operations made it possible to discover the tornado vortex signatures forming in them. At the time, NSSL did not have the capability of processing these data in real time (i.e., as it happened), but had to record the Doppler observations on computer tape and then analyze them later.

In a separate research project, the Air Force Cambridge Research Laboratory (AFCRL), outside Boston, was working on a method of visualizing the data from the Doppler radars in real time using a device called Plan Shear Indicators. The Plan Shear Indicators enabled meteorologists to visualize differences in wind speed along the radar's line of sight. Since the success of the Doppler radar system would benefit both civilian and military weather forecasting communities, a spirit of cooperation was in full force during this exciting time.

The National Weather Service borrowed one of the Air Force Plan Shear Indicators and installed it early in 1973. Shortly after the borrowed device had gone on line at Norman, Oklahoma, it proved its worth in a dramatic way. On May 24, 1973, Union City, Oklahoma, was hit by a devastating tornado. Forty-one minutes before the twister touched down, the Plan Shear Indicator had enabled NSSL meteorologists to detect the mesocyclone with the tornado forming in it. In a summary report written later, the reporter stated that it was now possible not only to recognize "a potentially threatening mesoscale cyclone aloft" but also "to monitor its development and descent" thereby ensuring accurate, short-term warnings of destructive winds.

NSSL's success in detecting tornadoes in 1973 was part of a far-reaching five-year research program on which it had

embarked in 1970. Not only were they able to "see" mesocyclones with tornadoes forming in them, but they were
also successful in following the fully formed tornadoes as
they descended to earth, did their damage, and then dissipated. An NSSL analysis of the project summarized in 1976
showed that more than 60% of the mesocyclones observed
by the Doppler radar produced tornadoes and the remaining 40% contained damaging winds and hail. Furthermore,
the report showed that the average lead time between the
first detection of the mesocyclone and the occurrence of
severe weather was often as much as 36 minutes; and that
the presence of these "tornado vortex signatures" (i.e.,
strong winds embedded within the mesocyclone) made it
possible to establish the location of tornadoes within a
kilometer.

The obvious success of the program had immediate
consequences. The program, after all, had been only experimental, and it had used experimental techniques and devices that might not do well under actual operational conditions. In a conference at Norman in September 1976, NOAA
scientists decided to begin a series of operational tests. The
tests, to take place during the next 5 years at Norman and at
Cimarron Field (also in Oklahoma), would evaluate the
adequacy of the real-time operational tests.

The test and development effort became known as the
Joint Doppler Operational Project (JDOP) and was seen by
NOAA as the key demonstration project that, if successful,
could be used to sell this new system to Congress. The
JDOP exercise began in 1977, with both the NWS and the
AWS providing the forecasters. The NWS brought in special
color displays to Norman to supplement NSSL's white-on-
black display. The Air Force Geophysics Laboratory (the
newly renamed AFCRL) brought in an engineering crew, the

latest in radar data processing display and recording equipment, and the most sophisticated color display to be found anywhere at the time. This was connected to the Cimarron Field radar. In addition, the radar operators at both locations had available to them for instruction and advice top NSSL and AFGL personnel who had participated in the Doppler experiment and developments at Norman and Boston in the late 1960s and early 1970s.

The radar operators were able to detect, sometimes at remarkably great distances away, the moving internal structures of thunderstorms and storm clouds and, thanks to automation and digitization of the radars, to analyze the small scale circulation patterns of the precipitation particles. As the mesocyclones developed, radar operators could "see" these strong winds using visualization techniques whereby strong motions in one direction were displayed in yellow and strong motions in the opposite direction were displayed in purple. The resulting patterns revealed what conventional radars could not: radar signatures of high wind shears, mesocyclones, and tornadoes. In addition, the operators knew when they saw evidence of strong outflows below the bases of clouds that they were looking at situations favorable to the creation of *gust fronts*, another potentially dangerous weather feature. And when they saw evidence of strong updrafts at cloud bases and tops, they also knew that they were dealing with the likely possibility of damaging hail.

The great success of the first year of operation in 1977 was followed by an equally good year in 1978 and by then it became clear beyond question that the future belonged to Dopplers. During both years, the JDOP forecasters, again and again, "saw" tornadoes developing within the mesocyclones as much as half an hour before they began descending to earth and doing damage. As a result, forecasters were able to issue tornado warnings that reached the tornado site a full 20 minutes or more ahead of conventional warnings from the NWS located in Oklahoma City. And that 20-minute lead time is all the more significant when one stops to realize that conventional warnings, derived from the conventional radars and confirmatory spotter information, had an average lead time of no more than two minutes before tornado touchdown.

The Doppler radars accomplished everything that could be expected of them. In addition to sending more than 200 accurate advisories to NWS field offices and Air Force bases, the system identified accurately, and on an up-to-the-minute basis, mesocyclones at distances of up to 250 kilometers and tornadoes at least half as distant.

The Federal Aviation Administration was so impressed with the test's accomplishments that, by the time the 1977 results were in, it formally joined the project and would become a full partner of the NWS and the AWS in support of the development and eventual operational deployment of NEXRAD. The FAA had other reasons for wanting to see NEXRAD become a reality.

On April 4, 1977, just as the first JDOP operation was getting started, a Southern Airways DC-9 ran into an area of severe thunderstorms at New Hope, Georgia. The airplane lost both engines in the storm. While its crew attempted an emergency landing on a state highway, the plane crashed, killing 62 people and injuring 22 others. As the National Transportation Safety Board was later to conclude, a major reason for the crash had been the inability of the FAA's air traffic control system to provide the DC-9 crew with reliable up-to-date information revealing the presence of the killer thunderstorm that the plane had run into. It did not take long for the FAA to realize the full significance of what had happened, and to see that a system like NEXRAD was the only answer to averting such disasters in the future. As a result, the agency, which previously had not concerned itself with the quality of the weather radar service it was receiving from the NWS, became one of NEXRAD's strongest backers. With such strong backing, the Office of Management and Budget (OMB) began developing appropriate budgetary justification for the new system. The first funds for NEXRAD were included in President Reagan's 1981 budget. The effort was coordinated through the Joint System Program Office (JSPO) and the Department of Commerce was designated the lead agency for the conversion of the outdated radars to the new generation of radars dubbed NEXRAD.

In the past, the problem was that technology outpaced its ability to assimilate the data it created. The JSPO realized this limitation and concentrated much of its effort on the automation of NEXRAD. This required the development of a great number of information-processing algorithms covering the known characteristics of all types of severe weather. For despite how good the technology that went into NEXRAD was, all the output from this sophisticated system was of little use if the forecaster in the NWS or AWS field office didn't know how to interpret the information. With the explosion of computer power gripping the world in the early 1980s, it became possible to develop the sophisticated software necessary to turn the pulses being transmitted by these new radars into information that could be pulsed over the communication airways to save scores of lives.

Working out the necessary software for the effort presented a double challenge. First, it required the development of computer algorithms to identify, track, and predict storm hazards, and second, the creation of display products to combine the algorithms' results with effective, computerized presentation of the data. The JSPO decided that the most effective way to develop what was needed was to let each of the participating agencies take the lead in one of the particular areas. The result was a NEXRAD system that is more than just a radar; it is a system with three distinct but integral components.

The radar itself is called the Radar Data Acquisition (RDA) Subsystem. This subsystem houses the powerful Doppler radar and is unmanned, designed for remote operation. Sitting atop a tower that rises high above the ground, the RDA contains the radar and also includes a transmitter so that its data can be sent somewhere for the actual analysis. Its responsibility will be to acquire the system's Base Data, the data having to do with reflectivity and atmospheric motion, and then send the data on for processing.

The second part of NEXRAD is the Radar Products Generation (RPG) Subsystem. This is the primary data processing part of the system. It will include computers, computer software, and communications hardware and software. It will be the host of all the meteorological algorithms used in the system's software and the source of all the products that will be called upon by the NEXRAD users. It will receive the Base Data from the RDA and generate real-time information on severe thunderstorms, tornadoes, flash floods, turbulence, gust fronts, and wind shear, and will have them ready to transmit to users in the form of basic graphic products that can be readily interpreted by the user.

The third component is the Principal User Processor (PUP) Subsystem, the set of color monitors manned by trained users in the field forecast offices. Communication between the RPGs and the PUPs is accomplished by high-speed telecommunication lines. The PUP subsystem allows the forecasters to tap into the store of radar products available to them at the RPGs, and to secure instantaneously whatever NEXRAD weather data they need. The systems underwent thorough testing through the mid-1980s and the operational unit being deployed into the field has been designated the WSR-88D.

As the 1990s progress, the NEXRAD units are being installed around the country at a cost of more than a billion dollars. By the time the entire network is in place (tentatively by 1997), it will consist of about 160 systems. The cost will have come out of the budgets of the NWS (~50%), the AWS (~30%), and the FAA (~20%). In his article that appeared in *Weatherwise* in 1986, Samuel Milner, a former historian of the Air Weather Service, perhaps best summarizes the impact of NEXRAD:

> However things turn out with the Dopplers, the fact remains that when the Doppler network becomes operational, it will have affected a revolution in radar meteorology—a revolution whose dimensions can perhaps best be illustrated by a famous poem by Christina Rossetti that some of us learned as children:
>
> > Who has seen the wind?
> > Neither you nor I.
> > But when the trees bow low their heads,
> > The wind is passing by.
>
> That no one had ever seen the wind was once the case. But no more. Now for the first time in history, Man can 'see' the wind as it works its mischief in the storm clouds. And armed with that knowledge, to a degree not hitherto possible he will be able, when the NEXRAD network becomes operational, to take informed and timely action to prevent that mischief from harming him.[23]

BEYOND NEXRAD: FILLING IN THE GAPS

As promising as NEXRAD appears, there is a problem. By themselves, the sites of the new radar system leave gigantic holes in any grid of the atmosphere, since they possess only 125 miles of tornado-detecting capability. To be effective, this would mean placing NEXRAD sites at 125-mile intervals across the Great Plains. The other problem is that they can't detect wind motions above about 2 miles of the earth's surface.

But part of NEXRAD's information gap can be filled by another development using Doppler radar: the wind profiler. The wind profiler can sense winds from about 1600 feet above ground to heights of 55,000 feet. A wind profiler works by pointing a beam of Doppler radar microwave vertically above itself. Then, by tilting the beam 15°, it measures wind components. By comparing these beam readings forecasters can determine upper-air wind speeds and direction, which would help predict storm tracks. Wind profilers would supplement data now gathered by the launching of weather balloons equipped with radiosondes. Currently, such balloons are launched twice a day from more than 250 sites in the United States and elsewhere. Wind profilers would be able to monitor upper-air wind conditions every hour, rather than every 12 hours. But wind profilers, when they become operational, won't replace balloons since the balloons monitor temperature and humidity also.

Another technology that is affecting positively our ability to forecast severe storms is a system of new computer algorithms that allow immense quantities of data to be condensed into convenient forms for display on desktop computer terminals. This system, called Automated Weather Information Processing Systems (AWIPS), allows a forecaster to use a mouse-driven control to call up just about any weather product. For example, a forecaster can sit at his terminal in Denver and with a click of the mouse call up a satellite view showing atmospheric moisture. With another click he can superimpose that onto a map of current jet stream activity. Without leaving his desk, then, he can interpret the latest data from satellites and ground-based sensors and use them for his local forecast.

The addition of such technology to ground stations, plus the new instruments aboard the new generation of satellites, promises to make the forecasting of severe storms and tornadoes more accurate and timely. By the early part of the next century, forecasters may be able to "see" tornadoes as they hide in the storm clouds, be able to track those storms and issue warnings well in advance of their arrival.

SEVEN

Long-Term Weather Prediction: Are Forecasters Full of Hot Air?

When it comes to what people think the future climate might bring, two well-known sayings reflecting opposite concerns for the future come to mind. The first is Chicken Little's classic warning, "The sky is falling! The sky is falling!" At the other extreme, one might argue, "What goes around comes around," reflecting the thought that nothing ever really changes, but that we are experiencing different sides of what are very natural cyclic variations.

Many people know the current sources of long-term weather forecasts. From the *Farmer's Almanac*, which forecasts the entire year's weather region by region, to the National Weather Service's 30-day general forecast that comes out every month, to the local weather forecaster who dares to predict what kind of winter or summer we'll be experiencing, today's public is often bombarded with forecasts that make it into our morning newspapers. As we indicated earlier, the use of computers and sophisticated numerical prediction models of the atmosphere have enabled forecasters to make more accurate one- and three-day forecasts. But what about our ability to predict long-range weather? In this chapter, we'll explore what scientists understand about variations of weather patterns from one year to the next and how this information may be used to produce long-range forecasts. And how does such a seasonal forecast fit into the theories regarding global climate change? These questions point to an even larger issue: With all our activities across the globe, are we disturbing this planet's climate? During the past two decades there have been so many conflicting theories and opinions among scientists, it is no wonder the public is confused. Let's start with what we know about the earth's climate and what possible changes may be taking place.

WHEN SCIENCE BECOMES AN ISSUE

Even as recently as the 1960s, the extent of our knowledge of climate could be summarized by a classification system that listed the climate of Miami as a "moist, tropical climate" and Minneapolis as a "temperate, continental climate."[24] The study of climate consisted of defining what kind of weather regime could best describe, on average, the prevailing weather at a particular location. But during the 1970s some scientists speculated that because of human activities such as the burning of fossil fuels and the paving over of vast tracts of land, the earth's climate was going to change. Other scientists didn't agree. Various theories and speculations of doomsday scenarios emerged like mushrooms after a rain. The result was public confusion. Were we doomed to destroy the earth by our greedy consumption of its resources, or was the earth very capable, thank you, of supporting the human race? Was the earth growing warmer, or colder? Was there a "greenhouse effect" or not?

Reid Bryson, a professor of meteorology at the University of Wisconsin, warned that, on the contrary, the earth's climate was going to be cooler. He reasoned that farming practices, such as planting and harvesting with large machinery, stirred up vast areas of topsoil. These particles of topsoil drifting around in the atmosphere would act as a reflective shield and keep the ground cooler.[25]

Syukuro Manabe, on the other hand, at the Geophysical Fluid Dynamics Laboratory in Princeton, New Jersey, held a contradictory view. It was Manabe's pioneering work that first called attention to the fact that observed increases in carbon dioxide would result in a general warming of the climate. This view gained ground during the 1970s and became a primary concern of various environmental groups in the 1980s. Such concern eventually gained worldwide acceptance and led to the establishment of the Intergovernmental Panel on Climate Change (IPCC) in 1988.

The IPCC was under the joint direction of the World Meteorological Organization (WMO) and the United Nations Environmental Program (UNEP). The IPCC's first order of business was to set up three working groups: one to assess the available scientific information on climate change; a second to assess environmental and socioeconomic impacts of climate change, and another to formulate response strategies. The first working group of the IPCC published a summary of its findings in 1990.[26] More than 170 scientists from 25 countries prepared the report. Before it was released, an additional 200 scientists also reviewed it. These numbers are quite extraordinary in the field of science, so it is fair to say that this effort was the most wide-reaching treatise in any scientific discipline ever produced. It was truly a global effort, in large part because it addressed an issue of global concern.

THE IPCC REPORT: TELLING IT LIKE IT IS

Most scientists will agree that weather forecasting isn't an exact science. Our accuracy rate decreases as the length of time we are forecasting for increases. The accuracy rate for a 24-hour forecast is around 80%, dropping to about 50% for anything over a three-day interval. (Imagine how inexact our prediction would be if we dared to forecast the weather a 100 years from now.) This inexactness has led to confusion in the public mind regarding just how valid weather predictions are. Facts, theories, speculations, and opinions have all been given their due in the press. The result has been a hodge-podge of conflicting scenarios. It was just these problems that the IPCC was mandated to clarify. Its role was to separate fact from fiction, consensus from individual speculation. The panel was to report on what was actual and what was not, what was a reasonable guess and what was wild speculation. The group divided its findings into three categories: what it was *certain* about; what it was *confident* about; and what it could confidently *predict* about the future weather.

According to the IPCC report, the committee was *certain* that

- A greenhouse effect exists; and
- Emissions resulting from human activities are substantially increasing the atmospheric concentrations of particular greenhouse gases [carbon dioxide, the

chlorofluorocarbons (CFCs), and nitrous oxide] and that these increases will enhance the greenhouse effect, resulting in an additional warming of the earth's surface.

Furthermore, the committee was *confident* that

• Some gases are potentially more powerful than others at changing climate, and their relative effectiveness can be estimated. Carbon dioxide has been responsible for over half of the increase in greenhouse effect in the past and is likely to remain so in the future.
• Atmospheric concentrations of the long-lived gases (carbon dioxide, nitrous oxide, and the CFCs) adjust slowly to changes in their emissions and that continued emissions at the present rates would commit us to increased concentrations in the centuries ahead. In other words, there is a lag in time between decreasing the source of these gases and any discernible decrease in their effect. The longer emissions continue to increase at present rates, the greater future reductions would have to be for concentrations to stabilize at a given level.
• To stabilize their concentrations at present-day levels, the long-lived gases would require immediate reduction in emissions from human activities of at least 60%.

Lastly, based on model results, the IPCC *predicted* that

• If emissions are not curtailed, global mean temperature during the next century will increase 0.5°F per decade with an uncertainty range 0.4 to 0.9°F; this is greater than that experienced over the past 10,000 years.

- Global mean temperature will be 2.5°F warmer by 2025 and 7°F warmer by the end of the next century.
- Land surfaces will warm more than the oceans.
- Climate changes will not be consistent. Certain regions will experience greater variability than others. For example, temperature increases in southern Europe and central North America are predicted to be higher than the global average while precipitation will be lower than average.
- Sea level will rise a bit over 2 inches per decade (with a range of 1 to 4 inches per decade), primarily because of thermal expansion of the oceans and the melting of some land ice. The report predicts sea levels may rise as much as 8 inches by the year 2030 and over 2 feet by the end of the next century.[27]

This predicted rise in sea level would create an entire new configuration of the United States and Europe. Entire countries would disappear (notably the Netherlands and parts of Belgium), while in the United States parts of Louisiana, most of Florida, and some of the major cities along the eastern seaboard would be under water. Previously landlocked cities would be transformed into ports, while the ports of Galveston and New Orleans would no longer be functional.

Scientists are more concerned about the enhancement of the greenhouse effect, that is, the rate of increase, rather than the greenhouse effect per se. The greenhouse effect is nothing new. In fact, without the greenhouse effect we would find our planet a very uncomfortable place to live. A schematic representation of the greenhouse effect is shown in Figure 23.

The greenhouse effect is important to us because the earth receives a fixed amount of radiation from the sun. This

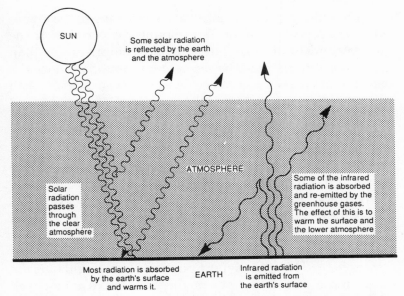

FIGURE 23. A simplified diagram illustrating the greenhouse effect. (From *Climate Change*, Intergovernmental Panel on Climatic Change, 1990, Cambridge University Press, Cambridge, U.K. Reprinted with permission.)

fixed amount of radiation is dependent only on how much radiation the sun puts out and how far the earth is from the sun. Based on these simple facts, the surface temperature of the earth should be 0°F. Actually, however, the average surface temperature is 59°F. The difference represents the difference between life as we know it and hardly any life at all, and is the result of the natural greenhouse effect of the atmosphere. Instead of all the heat intercepted by the surface of the earth being reradiated back to space, some of the molecules in the atmosphere absorb that radiation and then release it back to the surface, warming it.

The most important molecule contributing to the greenhouse effect is water vapor. The amount of water vapor in the air is extremely variable, but generally comprises less than 2% of the atmosphere. The second most abundant trace gas contribution to the greenhouse effect is carbon dioxide. Both of these gases are considered trace gases since compared with nitrogen and oxygen (which comprise 99% of the air) they exist in relatively small (or trace) amounts.

Furthermore, there is now scientific evidence buried in Antarctic ice cores that shows a very strong correlation between the amount of carbon dioxide and methane in the atmosphere and the average local temperature. Figure 24 shows the measurements from ice cores dating back as far as 160,000 years illustrating how the earth's temperature has closely paralleled the amount of these two trace gases in the atmosphere. The IPCC report points out that although the cause-and-effect mechanics of these variables is not well understood, calculations do indicate that part of the 9–12°F temperature changes observed over the past 160,000 years could be explained by the variations in the concentrations of these trace gases.

Despite the fact that these trace gases have shown a considerable amount of fluctuation over geologic time even without the large inputs from human activities, we do know that carbon dioxide and methane levels now are considerably higher than any measured from the ice core data. This suggests that anthropogenic activity has placed an unprecedented strain on the natural system.

For example, carbon dioxide concentrations fluctuated between 180 and 300 ppm (parts per million) over the past 160,000 years, yet concentrations in the 1990s now exceed 350 ppm. Similarly, methane varied between 300 and 700 ppb (parts per billion) over geologic time and now have more

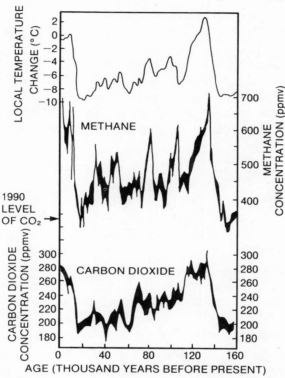

FIGURE 24. Concentrations of carbon dioxide, methane, and an analysis of temperature determined from Antarctic ice cores over the past 160,000 years. These plots show how global temperatures have been closely correlated with atmospheric concentrations of carbon dioxide and methane. It is also noteworthy that current concentrations of these trace gases are considerably higher than even the highest measured values over the past 160,000 years. (From *Climate Change*, Intergovernmental Panel on Climatic Change, 1990, Cambridge University Press, Cambridge, U.K. Reprinted with permission.)

than doubled to over 1700 ppb (or 1.7 ppm). Although the concentrations of these gases remained relatively steady up to the 18th century, more recent data (see Figure 25) suggest previously unheard-of increases since the dawn of the 20th century.

The concern is that the increases in these trace gases have been the primary reason for the observed increase in surface temperatures over the past century. Figure 26 shows the yearly average global temperature change from 1861 through 1989 relative to that for the 30-year period, 1951–

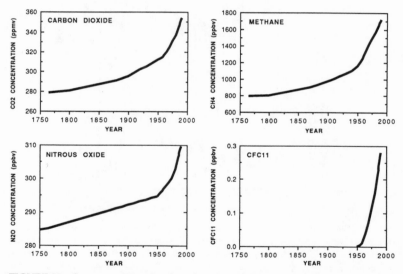

FIGURE 25. Concentrations of carbon dioxide, methane, nitrous oxide, and chlorofluorocarbon-11 between 1750 and 1990. After remaining relatively constant through the 19th century, these primary long-lived gases that trap infrared radiation have increased dramatically since the dawn of the 20th century. Chlorofluorocarbons were not present in the atmosphere prior to the 1930s. (From *Climate Change*, Intergovernmental Panel on Climatic Change, 1990, Cambridge University Press, Cambridge, U.K. Reprinted with permission.)

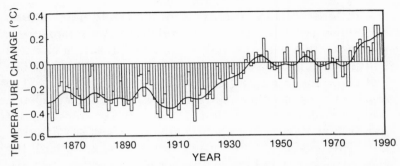

FIGURE 26. Observed global temperature change over the past 130 years. (From *Climate Change*, Intergovernmental Panel on Climatic Change, 1990, Cambridge University Press, Cambridge, U.K. Reprinted with permission.)

1980. The big question, of course, is whether the observed increase is a result of the increased loading of greenhouse gases in the atmosphere during the same period of time.

One scientist who believes strongly that the observed increase in carbon dioxide is contributing to the observed rise in temperature is Dr. James Hansen, the director of the Goddard Institute of Space Studies in New York. In fact, Hansen was so sure of his belief that he testified before a Congressional subcommittee on June 23, 1988, that he was "99% sure that the relatively high temperatures experienced in the United States during the 1980s were a sign that the greenhouse effect resulting from the higher levels of carbon dioxide is already upon us."[28] One year later, it was reported that the mild-mannered Hansen bet a colleague that one of the next three years would be the warmest of the century. He did not have to wait very long: an analysis of the temperature data for the next year confirmed that 1990 held the distinction as the warmest year of the century (up to that point).

In June 1991, an eruption from the Philippine volcano, Mt. Pinatubo, spewed more debris into the stratosphere than any other event of the century. As a result, a cloud of sulfuric acid formed in the stratosphere (more than 60,000 feet above the ground) that blocked out enough sunlight to result in a noticeable cooling at the earth's surface. Determinations showed that 1991 tied 1990 as the warmest year of the century despite the cooling effect of the volcanic eruption. Events such as the latter can continue to disturb the climate for years and make it impossible to formulate accurate predictions well into the future unless they are taken into account.

There are other forces that drive the climate from one year to the next, a field of study that scientists call interannual variability. It has emerged as one of the most fascinating aspects of weather and climate prediction.

THE CHRISTMAS CHILD
AND THE WEATHER HE BRINGS

At the turn of the century scientists noticed that atmospheric surface pressure fluctuations at Buenos Aires, Argentina, and Sydney, Australia, were out of phase with each other. Unlike surface pressure at middle latitudes, where variations are primarily determined by the passages of high- and low-pressure systems associated with the migration of frontal systems, pressure fluctuations in the tropics generally undergo very little daily variation (unless a tropical storm or hurricane is in the vicinity). What scientists in Buenos Aires and Sydney noticed, however, was that when pressure was relatively high over the course of several months at one location, it was relatively low at the other.

Further studies and an analysis of nearly 100 stations located at low latitudes showed that this oscillation had a period of between three and four years and that it was nearly global in extent. In a series of scientific papers published in the 1920s and 1930s, Sir Gilbert Walker, director-general of the observatories in India, referred to this large-scale phenomenon as the *southern oscillation*.[29] Although the pressure variations are most pronounced at southern tropical latitudes, the effects of the southern oscillation can also be observed over many other parts of the world.

Several decades later, in the 1960s, scientists discovered an important relationship between the southern oscillation and sea surface temperature variations in the tropics. They observed that high surface pressure over the western tropical Pacific Ocean (e.g., Australia) and low pressure over the eastern tropical Pacific Ocean (e.g., off the west coast of South America) coincide with heavy rainfall, unusually warm surface waters, and relatively weak trade winds in the central and eastern tropical Pacific.[30] This was one of the first geophysical observations that linked together the complex interaction between the ocean and the atmosphere. We now know this as *El Niño*.

The story of El Niño began to unfold in 1891 when Dr. Luis Caranza, president of the Lima Geographical Society, published a small article in the society's bulletin noting that a countercurrent flowing from north to south had been observed between the ports of Paita and Pacasmayo. The Paita sailors, who frequently navigate along the Peruvian coast in small boats, named this countercurrent the current of El Niño (the Child Jesus) because it had been observed to appear immediately after Christmas. The coast of Peru is usually a barren desert adjacent to a cold ocean that teems with fish and other forms of marine life. The warm current of the El Niño moderates the low sea temperatures during the early months of the year and is accompanied by heavy rains. This was the case in 1891 when it was observed that the temperatures along that portion of the coast were unusually warm. Observers thought the increased temperatures were a direct outcome of the hot current that "bathed the coast."

Such years were known as *años de abundancia*, years of abundance. In a report before the Sixth International Geographical Congress in Lima, Peru, in 1895, Señor Frederico Alfonso Pezet reported some of his observations: "First of all the desert becomes a garden when the soil is soaked with heavy downpours, and within a few weeks, the whole country is covered with abundant pasture. The natural increase of flocks is practically doubled and cotton can be grown in places where in other years vegetation seems impossible."[31] Other changes in the area included the appearance on the usually barren shores of Peru of long yellow and black water snakes, bananas, and coconuts, replacing the normal bird and marine life temporarily.

Although the impression persists that El Niño is a temporary departure from a generally "normal" condition of the tropical Pacific, this is not accurate. Normal conditions can be defined statistically, but such is not the case for this region of the world. In actuality, the southern oscillation either is just entering or leaving one side of its phase or is either just entering or leaving its complementary phase, for which the term *La Niña* has been applied. During La Niña, surface pressure is high over the eastern but low over the western tropical Pacific. During this phase of the southern oscillation, trade winds are generally intense and the sea surface temperatures and rainfall associated with it off the Peruvian coast are generally very low.

The terms El Niño and La Niña cover a wide range of

conditions, and are now classified into four categories: strong, moderate, weak, and very weak. Based on a quantification of the pressure difference measured at Tahiti (in the central Pacific) and Darwin (Australia), since 1937, "strong" El Niños have occurred in 1941, 1957, 1958, 1972, 1978, 1982, and 1986. But even within this classification of "strong" El Niño years, defined as years in which substantial pressure differences exist between the two sides of the Pacific, marked variations still arise. For example, the pressure differences between the two sites may last for nearly two years (as was the case in 1941), or for only a few months (as was the case in 1982).

TELECONNECTIONS

One of the most exciting findings in recent years in the field of meteorology is the effect of the southern oscillation on weather patterns at great distances from it. We have already seen how Bill Gray has used information about the southern oscillation to guide his forecasts for the upcoming hurricane season. Scientists have used the *teleconnections* to label this tendency of the southern oscillation to affect weather patterns in other parts of the world. As more and more data are scrutinized, scientists are just now starting to realize how complex the global weather machine is and how intertwined the atmosphere and the oceans are. Furthermore, we now realize that a complete understanding of both atmospheric circulation and ocean circulation is vitally needed to understand the general circulation of the atmosphere.

Scientists have recently discovered that the surface temperature of the water directly affects the precipitation patterns in the tropics. The most obvious shift occurs in the Pacific Ocean and can readily be observed from satellite images of cloud patterns. The bulk of the rainfall over the Pacific tends to be concentrated in two relatively narrow zones related to the convergence of the trade winds in the tropical Pacific. At about 8° to 15° N, the Intertropical Convergence Zone (ITCZ) comprises a west-to-east oriented band of rainfall extending across the entire Pacific from the Philippines to the Central American coast. South of the equator, the South Pacific Convergence Zone (SPCZ) is a somewhat broader region of heavy rains extending southeastward from New Guinea toward the Polynesian region. Unlike the ITCZ, the SPCZ is interrupted on its eastern side when cold water is present. Although both convergence zones exist throughout the year, they are notably stronger during their respective summer seasons.

During the El Niño years (i.e., when sea surface temperatures in the eastern Pacific are relatively warm), both the ITCZ and the SPCZ move closer to the equator, and even appear to become merged over their western portions near the international date line. This causes wetter than normal conditions along the equator, while unusually dry conditions persist in the usual positions of the convergence zones such as the islands of the western Pacific and the South Pacific islands of Fiji and New Caledonia. Farther west, drier than normal conditions affect a broad region of the tropics bordering the eastern Indian Ocean, such as Australia, Indonesia, and the monsoon that normally is present over southern Asia. It is noteworthy that some of the most extreme failures of the Indian monsoon have occurred during the "warm" (El Niño) years of the southern oscillation, whereas years of vast flooding over India generally coincide with the "cold" (La Niña) years. One exception to this general pattern, however, is that the island of Sri Lanka, southeast of India, actually experiences wetter than normal conditions during the warm El Niño years.

Over Africa, the southern oscillation affects a smaller percentage of the continent than it does across the Atlantic. During southern summer and fall (i.e., January through April), there is a strong tendency for precipitation to be greater than normal during the onset of an El Niño. This phase of the southern oscillation affects a rather large region in southeastern Africa that includes Zimbabwe, Mozambique, and South Africa. However, once the El Niño has already been established, this same region often experiences widespread drought. By some accounts, the drought in southern Africa in 1992 was the worst of the century and it followed the moderate, but prolonged, El Niño that persisted through 1991 and 1992.

The highlands of Ethiopia and Somalia also generally experience drier than normal conditions during an El Niño year. There is also some evidence that El Niño years coincide with dry conditions over the Sahel region bordering the Sahara desert north of the equator.

The strongest precipitation-related teleconnection in North America seems to be found along the Gulf Coast region of the United States and parts of northern Mexico, Texas, and the Caribbean Islands. During El Niño years, wetter than normal conditions are generally observed during the winter months. This signal over the southeastern United States is one of the most consistent extratropical teleconnections associated with the southern oscillation and was evident to Sir Gilbert Walker in the early 1920s. More recently, scientists have noted that when winters are wetter than normal in the southeastern United States, wet conditions often prevail over the southern High Plains. Both of these findings seem to be associated with a strengthening of the jet stream over the Gulf of Mexico during warm El Niño years, which tends to bring about a more active storm track that affects these regions in the United States. In Hawaii, on the other hand, winter rainfall is significantly diminished during El Niño years.

The most impressive teleconnection associated with the southern oscillation is the strong tendency for the global tropics as a whole to be warmer during El Niño years and colder during La Niña years. The effects are widely observed in central and southern Africa, eastern India, all of Southeast Asia, northern South America, along the entire west coast of South America, and in northeastern Australia. Outside of the tropics, warm winters in Alaska and throughout central and southern Canada also seem to be present in years following an El Niño. Colder than normal winters have been observed in Scandinavia and the southeastern United States during El Niño years. Thus, the Christmas Child first identified in the waters of the eastern South Pacific off the coast of Peru has now been implicated as the mischievous little being that also enhances the likelihood of droughts in Africa and winter flooding in the southeastern United States.

USING TELECONNECTIONS TO INTERPRET THE PAST AND FORECAST THE FUTURE

Hydrology is the science that studies the water cycle on this planet. A major subdiscipline of hydrology is the study of river and stream flow. The occurrence of droughts and floods over a particular region is what determines how much water flows through a particular river. Conversely, if we know the amount of water flowing through that river during a given year we can determine the rainfall over the region that eventually drains into that river. An examination of the data available from early in the 19th century through the present indicates that a strong or a very strong El Niño was present during the years in which water flow through the Nile River was statistically far below normal. With the presence of such a strong correlation, William H. Quinn of Oregon State University[16] has examined Nile River flow to determine whether or not the phenomenon of the southern oscillation can be detected over the past millennia.[32]

Droughts, floods, plagues, and famines have burdened northern Africa over the course of many centuries. Fortunately, there are records of many of these natural catastrophes, and they include fairly detailed data quantifying the water flow through the Nile River. The yearly flood of the Nile has been the basis of Egyptian agriculture for thousands of years. Before the dams and other regulatory facilities were built upstream, the river's flow level at Cairo was carefully recorded. Scientists can use this information as a surrogate to determine regional flooding and drought conditions and the relationship between precipitation in this region and the El Niño–southern oscillation.

Known history of the Nile dates back to just before 5000 B.C. The record of the Nile levels dates back to 3000 to 3500 B.C., when the river gauge had been called a Nilometer.[33] Three types of Nilometers were apparently used in these ancient times. The first type consisted simply of marking the water level on cliffs on the banks of the river; the second type consisted of a scale, usually made of marble, on which the water level was read; the third type brought water of the Nile into a well whereupon the water level was marked either on the wall of the well or on a central pillar.

Joseph is said to have built the first Nilometer at Bedre-shen on the west bank of the Nile near the remains of a wall said to be part of the "granary of Joseph." Since the Nile appeared to rise in flood at about the same time every year, its behavior might be described as regular. Nevertheless, the Bible tells us, ". . . there came seven years of great plenty throughout the land of Egypt. And there shall rise after them seven years of famine. . . ."[34] The interpretation by Joseph to this dream of the Pharaoh of Egypt was probably the first indication of interannual variability, but in this case this variability was foretold by Joseph to last seven years. Scientists speculate that the seven-year famine foreseen by Joseph may have in fact occurred around 1708 B.C.

The Moslem era in Egypt began about 622 A.D. The Arab conquerors maintained records with Nilometers in-stalled in Memphis and later at the southern end of Roda Island. Records on the minimum and maximum water levels each year were chronicled for the period 641–1469 A.D. Researchers discovered another set of data covering the years from 1504. The missing years of data, 1470–1503, have been described elsewhere as "years of plenitude," and thus we can infer no disastrous droughts occurred. A large data gap exists between 1523 and 1586, following the Turkish conquest of Egypt.

From such records scientists have determined that over a stretch of 900 years, 178 of those years were spent under relative drought conditions, when the river flow was weak. That's a frequency of about one in every five years. This frequency is slightly less than what has been observed with more reliable modern data which have been used to deter-mine that a strong or very strong El Niño has occurred every 3.8 years. Within those 900 years, however, there are certain periods that show significant differences. For example, the period 1694–1899 is often referred to as the Little Ice Age. During this time, the climate in the northern hemisphere was known to have been colder than normal and the Nile flow data show that droughts occurred every 2.8 years. Thus, it is likely that strong El Niño events were considerably more frequent during the time when the Little Ice Age gripped the planet. During this era, a seven-year period (1790–1797) of drought dried up the Nile. Thus, it is possible that seven years of famine could have existed if Joseph had been living in a relatively cool period in the year 1700 B.C.

According to the Nile data, the lowest river flow was measured in the year 1200 A.D.[35] It is quite likely that the very strong El Niño that may have been responsible for low river flow that year in northern Africa was also responsible for the cataclysmic "Chimu flood" in coastal Peru in the same year. In summary, the southern oscillation and the El Niños and La Niñas that define which part of the cycle the ocean–atmosphere system is in may have had profound effects on events that have influenced the weather worldwide in different ways. It is likely that the teleconnections brought about by the presence of the southern oscillation has often generated extreme weather events far removed in space, but closely related dynamically.

In general, when people hear of a global warming trend of one or even several degrees, it's difficult for them to grasp what effect on their lives, if any, such a small perturbation may have. There is little doubt that the human race can and will adjust to such climate changes. But the analysis of the Nile records suggests that throughout recorded history there have been prolonged periods (time scales of decades to centuries) of relatively cooler or warmer climates. During the relatively cool periods, the historical records suggest that droughts in northern Africa are likely to occur more frequently—even to the extent of being present for as long as a decade. So the question arises: If the planet does warm appreciably because of increased carbon dioxide concentrations, what kind of changes in the weather can we expect to see and where will these changes be most severe?

At the National Center for Atmospheric Research in Boulder, Colorado, researchers are trying to answer this question using their general circulation model. In this study, the background concentration of carbon dioxide is arbitrarily doubled everywhere in the atmosphere. The doubling of CO_2 results in a fairly uniform warming of the atmosphere on the order of about 2°F.[36] What this study showed was that the frequency of El Niño years has remained approximately the same (over the 15-year period that the model was run), but that the response to the El Niños in all latitudes of the northern hemisphere is more pronounced when the CO_2 is doubled relative to the currently observed concentrations of CO_2. In the southern hemisphere, the response to the El Niño under doubled CO_2 concentrations is only apparent at high southern latitudes (i.e., latitudes near the pole). Thus, the study indicates that the departures from "normal" conditions that often accompany El Niños in the northern hemisphere may become even greater as the overall temperature of the atmosphere increases with rising CO_2 levels.

Because the model used at NCAR must make some simplifying approximations so that it can be programmed and run on the Cray computers at the center, the accuracy of the prediction is still in question. But as scientists continue to unravel more of the historical data that are available, they will be able to define with a greater degree of accuracy exactly how well these models simulate weather patterns on the time scales of decades to years. Despite our progress in this field, there is still a considerable way to go.

PREDICTING THE FUTURE:
HOW MANY UNKNOWNS DON'T WE KNOW ABOUT?

When scientists issued the first warnings of global warming less than two decades ago, the finger was pointed at the observed increase in carbon dioxide as the primary culprit. Before the 1990 Intergovernmental Panel on Climate Change (IPCC) report, scientists tried to assign values to the various gases according to their contribution to atmospheric warming. In this way they hoped to pinpoint which gases were causing the most warming. In 1980, the general consensus was that the increased warming that had occurred since preindustrial times from higher concentrations of greenhouse gases was caused primarily by carbon dioxide, which accounted for about 61% of the warming effect. The values assigned to other gases were methane, 17%; chlorofluorocarbons, 12%; and nitrous oxide, 4%.

Because the methane present in the lower atmosphere gradually drifts to the stratosphere, where it undergoes a series of photochemical reactions eventually making water vapor, the resultant increase in stratospheric water vapor also accounts for 6% of the warming since the onset of the industrial era. However, by the end of the 1980s, the IPCC report estimated the contributions from the trace gases to be carbon dioxide, 56%; methane, 11%; nitrous oxide, 6%; chlorofluorocarbons, 24%; and increases in stratospheric water, 4%.[37] Thus, it appears that carbon dioxide plays less of a dominant role in the warming calculations as civilization dumps more and more new chemicals into the atmosphere.

Even since the issuance of the IPCC report, scientists have come to realize that there are other extremely important pieces of the puzzle that likely affect future climate scenarios. One of these missing pieces is the identification of vast regions of the world from which particulate matter, such as ash from coal fires and wood burning, released into the atmosphere may have considerable climatic implications. A paper published by Robert Charlson and his colleagues at the University of Washington suggests that the presence of particulate matter over industrialized regions of the northern hemisphere has probably resulted in a widespread cooling of the atmosphere.[38] Because these particles reflect incoming sunlight, they tend to cool the atmosphere. Some scientists estimate that the net cooling effect of these particulates may be comparable to the warming that has been caused by the observed increase in carbon dioxide.

Carbon dioxide, however, is a fairly long-lived trace gas that circulates throughout the entire planet. Particulates in the lower atmosphere are not sufficiently long-lived to be spread uniformly across the planet, but they do build up in preferred regions close to where they are generated. Using calculations generated by a model of the global circulation and a knowledge of the sources of these particles, some scientists now believe that the presence of these particles should have counterbalanced the global warming by reflecting incoming sunlight. Most of the particulate matter is believed to be composed of sulfate particles resulting from the combustion of coal. Ironically, as air pollution technologies improve and as they are implemented in more countries around the world, the cleaner air may result in an increase in global warming.

One factor not addressed by the IPCC report is that of the observed increase in tropospheric ozone, i.e., ozone produced near the surface of the earth by photochemical smog. The IPCC scientists estimate that the contribution from increases in tropospheric ozone may have caused 10% of the warming since preindustrial time, but they also cautioned that such an estimate is extremely difficult to quantify because of ozone's short-lived nature. Tropospheric ozone lasts no more than a few days if it is present in the lowest few thousand feet of the atmosphere, but it may hang around for a few months if present in the middle or upper troposphere to altitudes of ~40,000 feet. If the ozone we generate as photochemical smog had a longer lifetime in the lower atmosphere, it would eventually drift to the stratosphere and help counterbalance some of the effects of the depletion of the ozone layer in the stratosphere. But the ozone pollution in the lower atmosphere does not hang around long enough to make it to the stratosphere since transport to the stratosphere requires a year or more.

This relatively short lifetime of tropospheric ozone is what makes its impact on the global climate so difficult to quantify. Because its distribution is so varied, a global increase in tropospheric ozone cannot be easily quantified from readings taken at a few surface stations at only several locations around the world. We get a much better picture from space of the pattern of tropospheric ozone distribution.[39] In Figure 27, we can see that tropospheric ozone emanates from each of the industrialized continents in the northern hemisphere and is blown downwind by the "prevailing westerlies." We can also see a large plume at low latitudes centered over the tropical South Atlantic Ocean.

These tropical high smog concentrations originate from both South America and Africa and result from the widespread practice of burning crops and vegetation. Despite a general misconception to the contrary, only a small percentage of the emissions released into the atmosphere come from deforestation. Most of the burning in these regions results from farmers and nomads burning vegetation and savanna to clear the way for crops. We can see from the satellite data

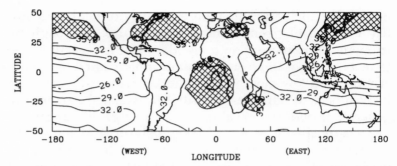

FIGURE 27. Tropospheric residual ozone (1979–1990): the global distribution of ozone in the lower atmosphere. This depiction has been derived from a technique using two satellite data sets. Of particular significance are the plumes coming from North America, Europe, Asia, and Africa. The pollution from the northern hemisphere continents is the result of industrialized combustion, whereas that from Africa is derived from the human activity of vegetation burning.

that the smog pollution from burning in Africa is about three times more extensive than that in South America, where there is so much debate over the destruction of the Amazon rain forest.

The plumes of ozone we see from space trailing off the coasts of North America, Europe, and Asia very likely did not exist before the turn of the century. All combustion processes, from a campfire to a steel-making inferno, can lead to the formation of photochemical smog and ozone. But one of the real champions of smog production is the internal combustion engine, most ubiquitously found in the autos of the industrialized nations of the northern hemisphere. Because of the high temperatures created inside the combustion chamber, the pollutants released by the internal combustion engine are ideally suited to react in the atmosphere to produce photochemical smog and the ozone that accompanies it. Thus, it is fair to say that tropospheric ozone has increased considerably during this century in regions that have become highly industrialized.

Two German scientists, Andreas Volz and Dieter Kley from one of the Energy and Atmospheric Research Institutes in Jülich, have studied just such a relationship, to determine whether or not industrialization has actually resulted in an increase in tropospheric ozone concentrations. They were helped by the existence of a remarkable set of records. Between 1876 and 1910, the French scientist Soret made detailed measurements of ozone at the Montsouris Observatory outside Paris. By today's standards, his method of measurement was fairly crude, but at the time it was the only reliable way to measure the newly discovered atmospheric gas. The Paris measurements were reported in increments called "Schönbein units," named for the discoverer of ozone, C. F. Schönbein. A "Schönbein unit" referred to a scale that related to how much a piece of paper soaked in potassium iodide changed color. Since potassium iodide reacted with ozone, the more ozone in the air, the deeper blue the paper turned.

In a paper published in the British journal *Nature* in 1988, Volz and Kley described how they carefully reconstructed the instrument used in the French Observatory and then recalibrated it using today's calibration standards.[40] Figure 28 shows a comparison of the ozone measurements from Paris with data from the northeastern United States and Germany approximately 100 years later. The increase of more than a factor of three over the past century is obvious and the fact that the highest surface ozone concentrations are now seen during the summer in the northern hemisphere also supports the hypothesis that the high values of today are a result of widespread generation of photochemical smog, which is most intense in the summer months.

But what does this mean? It is difficult to extrapolate

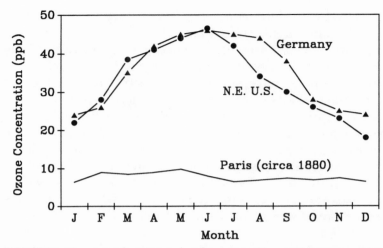

FIGURE 28. The seasonal variation of surface ozone. The monthly average values of ozone at locations in the United States and Europe in the 1970s and 1980s are compared with concentrations measured near Paris a century earlier. These plots show that surface ozone has increased dramatically over the past 100 years. In addition, the fact that the highest concentrations are now seen in the summer is further evidence that the observed increase is a result of local pollution. The natural source of ozone being brought down from the stratosphere should be most pronounced in the late winter and early spring, in agreement with the seasonal cycle observed in Paris circa 1880.

these measurements to the entire planet so as to quantify the impact of increases in tropospheric ozone on global climate. Tropospheric ozone does not have a long lifetime, and since we only have records from Europe and the eastern United States in this depiction, we should expect the increase to be the greatest at the two highly industrialized locations. On the other hand, even data from remote sites in Alaska and Hawaii show that tropospheric ozone has increased 0.8% since the mid-1970s.[41] Since that time, nearly every station in the industrialized world (where hundreds of stations have been monitoring surface ozone since the 1970s) shows an increase of 1–2% per year. Thus, it is indeed likely that surface ozone concentrations have risen nearly everywhere in the world. How much of an increase has taken place since the last century is a matter of speculation. But the limited amount of data do not rule out a two- to threefold increase in tropospheric ozone over the past 100 years.

The impact of increased concentrations of tropospheric ozone on climate is shown in Figure 29. This chart compares the radiative forcing (i.e., warming) by each of the trace gases with a hypothetical doubling or tripling of tropospheric ozone over the same period of time. During that same period, carbon dioxide concentrations have increased by 23% and methane by 43%. Because these two trace gases are so long-lived, we can be fairly confident that any trend observed at one location would be representative of its global increase.

The problem is further complicated because the impact of an increase in tropospheric ozone depends on the altitude at which the ozone is increasing. Up to altitudes of ~40,000 feet, the higher the ozone is situated in the atmosphere, the more effective it is as an absorber of the radiation responsible for global warming. Therefore, increased concentrations of ozone at altitudes of 30,000 to 40,000 feet are considerably more problematic than increased concentrations near the surface. The calculations shown in Figure 29 assume that ozone has doubled or tripled throughout the entire tropospheric column. Although the data in Figure 28 show the substantial increase over time at the surface, it is a leap of faith to extrapolate this to being representative of what is taking place at altitudes of 30,000 or 40,000 feet, where the increase in tropospheric ozone has a more pro-

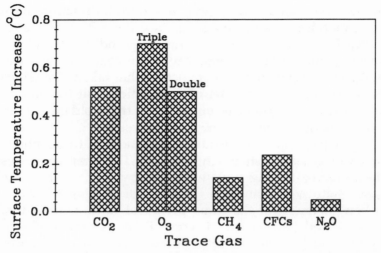

FIGURE 29. Contribution of greenhouse gases to global warming since the industrial era: calculated impacts of various trace gases since 1880. The bars show the calculated amount of warming related to the trapping of infrared radiation by various trace gases. The amount of warming of carbon dioxide (CO_2), methane (CH_4), the chlorofluorocarbons (CFCs), and nitrous oxide (N_2O) are calculated using the increases shown in Figure 25. The two bars for ozone (O_3) show the warming impact of increasing tropospheric ozone by factors of two and three over the same period of time using the same computer model of radiative transfer.

found effect on the earth's climate.[42] On the other hand, however, internal combustion engines on jet airplanes have been spewing forth smog precursors at these sensitive altitudes for several decades. Since the 1960s, measurements have been made at higher altitudes using ozone sensors on weather balloons. Their readings, which are considerably fewer in number than those observations at the surface and also probably less accurate, do suggest that ozone concentrations have increased throughout the entire troposphere during the past several decades, although the observed increase appears to be less than what has been observed at the surface.[43]

The potential importance of atmospheric particulates and tropospheric ozone, both the products of human activities, had not been considered in detail when the IPCC issued its report in 1990. It is only now being recognized that they may be making an enormous impact on perturbing the earth's climate. Still, as this book goes to press, the jury is out. There may be other findings in the scientific community in the next several years that may likewise alter our thinking about what are the most critical factors that are disturbing the earth–climate system.

What makes the study of climate so interesting is the unexpected. On June 12, 1991, Mt. Pinatubo, in the Philippine Islands, erupted violently over a week-long period. The eruption caused widespread devastation as volcanic ash covered many parts of the island, killing more than 700 persons and forcing more than 100,000 to lose their homes. The damage from the ash fallout to the Clark Air Force Base was so extensive that U.S. authorities decided it was cheaper to close the facility than to clean it up. Along with the ash, many trace gases were released into the atmosphere. Officials estimated that more than 20 million tons of sulfur dioxide gas were spewed into the stratosphere (altitudes of 50,000 to 60,000 feet) by the eruption.

Once in the stratosphere, sulfur dioxide undergoes a series of chemical reactions that eventually result in the formation of particulate matter composed initially of the volcanic ash and smaller particles that have been generated from sulfuric acid. Because of their relatively large size, the ash particles are affected by gravity and drift downward into the troposphere, where eventually rain will remove them. But the smaller particles resulting from the emissions of sulfur dioxide can remain in the stratosphere for several years.

The Mt. Pinatubo eruption is believed to be the largest such volcanic eruption this century. Unlike previous major volcanic eruptions, scientists are able to monitor the spread of the ash and the stratospheric sulfuric acid cloud thanks to the satellites that are now orbiting the earth, allowing them to quantify the effect of a major volcanic eruption on the earth's climate for the first time. Scientists have long suspected that major volcanic eruptions act to cool the earth. Following the eruption of Krakatoa in 1883, reports in many parts of the world described the "year without a summer." In northern New England, for example, below freezing temperatures were recorded every month, including July and August.

Even as recently as 1982, following the eruption of the Mexican volcano El Chichón, satellites were not in place to monitor the spread of the sulfuric acid cloud or to quantify the impact of this cloud on the earth's climate. But in 1984, instruments were placed on several satellites with the explicit purpose of monitoring the radiation budget of the atmosphere as part of ERBE, the Earth Radiation Budget Experiment. Using the data obtained from these satellites, Patrick Minnis, a research scientist at the NASA Langley Research Center in Hampton, Virginia, and his colleagues were actually able to calculate the effects of a major volcanic eruption on the climate for the first time.[44]

The ERBE is able to measure on a global basis how much sunlight is being reflected into space by clouds, land surfaces, and particles suspended in the atmosphere. The experiment also has the capability to determine how much heat is retained by the atmosphere as a result of the presence of these particles and clouds and other trace gases that contribute to the greenhouse effect. By comparing the data from August through October 1991 with the satellite measurements from previous years, Minnis and his colleagues were able to determine that the eruption of Mt. Pinatubo resulted in the earth uniformly cooling by 0.9°F.

The scientists were also able to track the dissemination of the ash cloud. In June, immediately following the eruption, the ash cloud was centered in the latitude band between 10° and 20° N, consistent with the latitude where the eruption took place (15° N). Scientists use the term *optical depth* to quantify how much light is actually blocked out or scattered by the aerosol layer; by August, the optical depth of the Mt. Pinatubo cloud was nearly three times greater than what had been observed in late June. Furthermore, the cloud had spread to a much wider belt between 40° N and 40° S. This cloud, which was mostly a result of the particles generated by the release of sulfur dioxide and its subsequent conversion to sulfuric acid, was now centered over the southern tropics at a latitude of 10° S because of the patterns of the winds that existed while the cloud was being generated. The remnants from Mt. Pinatubo should be observable through the mid-1990s.

Because of the cooling from the Mt. Pinatubo eruption, there has been a slowdown of the warming trend that has generally been observed since the 1980s. No matter how sophisticated or how complicated our numerical models of the atmosphere are, or may become, there will always be events such as volcanic eruptions that cannot be predicted. The possibility always exists that the occurrence of such an event may be overlooked in the development of scientific models. We have learned our lesson from the models that were used to predict ozone depletion in the stratosphere. At the time a substantial complement of scientists argued that there was no effect from the release of chlorofluorocarbons because the data in the late 1970s showed that ozone in the stratosphere was actually increasing. As we discovered later, one of the reasons for this increase was that the stratosphere was in the process of recovering from the losses that had occurred as a result of atmospheric nuclear testing in the late 1950s and 1960s. By the early 1980s, when this recovery process was coming to an end, and when enough chlorofluorocarbons had drifted to the stratosphere to start causing ozone depletion, scientists discovered that their models had completely neglected the mechanism that may have been most important in the destruction of the ozone layer. The original theories describing the destruction of the ozone layer by man-made chlorine-containing chemicals calculated that the most vigorous region of ozone destruction should be taking place in the upper stratosphere, at altitudes above 100,000 feet. When scientists discovered that the altitude of maximum ozone destruction was much lower in the stratosphere (around 70,000 feet), they also discovered that the primary reason for this destruction was by a mechanism they had totally ignored in their theory.

Researching the Antarctic ozone hole, scientists discovered that the regions of the most severe destruction were also regions of high thin clouds that became known as *polar stratospheric clouds*. These clouds form only in the coldest regions of the atmosphere. Above Antarctica the air in the lower stratosphere reaches the lowest temperatures ever measured in the atmosphere: below $-130°F$ ($-95°C$). The presence of these clouds greatly enhances the rate of ozone destruction through chemical processes that were completely unknown when the first theories of ozone depletion came forth in the late 1970s. It was a humiliating experience for atmospheric scientists, a reminder that the complex system of global climate is yet to be fully understood. And for this reason, there surely will be discoveries as new data sets are analyzed that will likely cause us to reevaluate our current forecasts about global change.

EIGHT

A "Hole" New Way of Forecasting

Consider the following for a science fiction movie plot: Big-time chemical company invents a harmless gas that helps keep together other chemicals so they can be put into a can and sprayed out under pressure. Other manufacturing companies love the stuff, for now they can put their foot powder, deodorant, hair lotion, whipping cream, house paint, lubricants, just about anything, into a can and the consumer just has to aim and press the nozzle to get a dose of it. Press once, and a jet of deodorant hits it mark. A simple spray on the hair and hair stays in place all day. Convenience.

But here's where the plot thickens. It turns out that this harmless gas used to propel other stuff out of a can is not so harmless, that it has the potential to threaten all life on earth. Absurd, right? That's just about how seriously many scientists and lay people accepted the hypothesis put forth by two chemists when they first published their scenario in the British journal *Nature* back in 1974.

The two chemists who released their report were not the villains of some science fiction film. They were indeed two concerned, diligent scientists: F. Sherwood Rowland (known to almost everyone as "Sherry") and his postdoctoral co-worker, Mario Molina. They were both at the University of California at Irvine when they submitted their paper, which theorized that although the propellants used in aerosol spray cans were harmless and inert, they were so inert that they were dangerous—they could never be removed from the atmosphere by any normal removal process.

Before the report was released, the fact that this propellant was so inert was assumed to be auspicious news, since most atmospheric removal processes of trace gases in the atmosphere ended up forming either acid rain or photochemical smog. Here was a chemical propellant that was nontoxic, nonreactive, and nonpolluting. All it did was drift ever so slowly up to the upper atmosphere, twice as high as commercial planes flew, above the earth's protective ozone layer. A good place for it, out of the way, in the earth's attic, so to speak.

That was what everyone thought until Rowland and Molina did some research on the subject and published their paper. What they found was that way up there in the upper atmosphere, the sun's intense radiation could provide enough energy to break the molecules apart into their basic elemental components. Once this happened, a series of chemical reactions in the stratosphere could take place and destroy ozone with incredible efficiency.

This process took a long time to begin, between 20 and 50 years for the molecules in the propellant to drift up to the upper stratosphere. But once it began, the resulting chemical sequence would eventually reduce the amount of ozone in the ozone layer to such an extent that its ability to protect life from harmful ultraviolet radiation would be greatly reduced.

The chemicals that Rowland and Molina were concerned about were chlorofluorocarbons (CFCs). They were used primarily as propellants in spray cans, for just about everything from deodorants to house paint. By the 1960s just about any product that previously was squeezed, pumped, spread, or poured could now be sprayed, thanks to CFCs. These chemicals were also used as the gas in air-conditioning compressors and therefore became omnipresent as air-conditioning systems became standard equipment in homes, offices, and cars.

As their name readily suggests, chlorofluorocarbons contain atoms of chlorine, fluorine, and carbon. When exposed to the high-energy radiation in the upper stratosphere, CFCs break apart, the atoms of chlorine, fluorine, and carbon going their separate ways. The fluorine and carbon atoms eventually react to form relatively harmless gases that disperse throughout the upper atmosphere. But the chlorine atoms engage in activity not so harmless. Every chlorine atom that is carried along by the CFC molecule up to the ozone-rich upper stratosphere can destroy up to 100,000 ozone molecules before the chlorine combines with an atom or molecule other than ozone to form another relatively benign molecule in the stratosphere.

The story of the ozone controversy is now well-documented, but at the time of that article, relatively few people—including many atmospheric scientists—realized that Rowland and Molina's warnings were much more than theoretical hype. In our earlier book, *Global Alert: The Ozone Pollution Crisis* (Plenum Press, 1990), we noted that the observations of ozone at the time of the Molina–Rowland paper did not suggest any depletion of the ozone layer. As other scientists went back through all the historical records of ozone measurements, no evidence could be found to support the contention that the ozone layer was starting to erode. These historical records included not only scattered systematic measurements (i.e., measurements using the same measurement technique over a long period of time) dating back to the 1920s but also a more standardized and carefully calibrated record of measurements begun in the 1950s.

Since the International Geophysical Year (IGY) in 1957, scientists expanded the number of locations where ozone measurements were made and took extreme care to ensure that these stations around the world used the same calibration standards so they would be able to compare results directly. As long as a universal calibration standard was used for every ozone measuring device everywhere in the world, scientists would not need to be concerned about potential differences resulting from slightly different measurement techniques.

So it was thought that a careful examination of the existing records at the time Molina and Rowland published their paper would show whether or not there was reason for concern. But instead of confirming the Molina–Rowland hypothesis, some of the papers published in the late 1970s suggested just the opposite: that stratospheric ozone was *increasing*. At the time, it certainly looked as though these two California scientists were crying wolf.

But in 1985, a dramatically new and unexpected set of measurements, also published in *Nature*, verified that the warnings first issued in the Molina–Rowland paper should be heeded. Scientists from the British Antarctic Survey reported discovering a "hole" in the ozone layer over Antarctica as the polar night ended in September and October. The British team, led by Joe Farman, described a series of measurements using spectrophotometers, instruments that determine how much ozone is in the atmosphere by carefully examining how much light at specific wavelengths is reaching the earth's surface. Since ozone is the only molecule that blocks out sunlight at certain specific wavelengths (carefully determined in controlled laboratory experiments), scientists can use this information to determine how much ozone must be present between the sun and the instrument's location (at the earth's surface). After considerable checking and rechecking every aspect of his findings, Farman finally convinced himself that what he had measured over the past several Antarctic springtimes was not a mistake: his sensors showed that ozone levels during the early 1980s were barely half of what had been considered "normal" values during the 1950s and 1960s at this particular time of the year.[45]

In 1986, with a new sense of urgency brought about by Farman's findings, the international scientific community convened a panel to assess all the recent findings that now seemed to support the Molina–Rowland hypothesis. Furthermore, if a decrease of ozone in the stratosphere could be definitively established, this panel was charged with the responsibility of finding out whether the decrease could be attributed to natural causes or human activity. Prior to the establishment of this fact-finding panel, several scientific papers, in addition to Farman's paper, had been published reporting that ozone amounts had been decreasing in the late 1970s and early 1980s, contrary to the findings that came out immediately after the Molina–Rowland 1974 paper. According to these studies, there were several possible explanations for the observed decreasing trend.

One distinct possibility was that one of the satellite instruments measuring continuously since 1978 had been drifting in its calibration. This instrument, the Total Ozone Mapping Spectrometer (TOMS), had already been operating flawlessly for more than seven years when the panel decided to examine whether or not the absolute numbers it had been sending back to earth were lower because of some kind of electronic drift of its sensor. After all, the instrument was supposed to be functional for only two years and it had already more than tripled its expected operational lifetime.[46]

So the panel was given the responsibility of reexamining the TOMS data and comparing these data with other ozone measurements from ground-based instrumentation.

Another possibility that would explain the observed decrease of total ozone during this period was a phenomenon called *solar variability*.[47] The sun undergoes an 11-year cycle during which the number of observed sunspots varies. Sunspots are storms on the surface of the sun. One manifestation of these storms is the emission of excess radiation that often occurs during periods of high sunspot activity. During these occurrences, the excess electromagnetic radiation emitted from the sunspots intercepts the earth's magnetic field, causing a disruption of radio communication on earth and sometimes a display of auroras at latitudes where they are not normally seen. As these energetic particles enter the earth's atmosphere, they interact with nitrogen and oxygen molecules in the upper atmosphere. The intense energy from these particles is so strong that these normally stable molecules can be split apart into nitrogen and oxygen atoms, where they can recombine in the rarefied upper atmosphere to form a molecule called nitric oxide. [Nitric oxide (chemical symbol, NO) should not be confused with *nitrous* oxide (N_2O), commonly referred to as laughing gas. This nitric oxide present in the upper stratosphere could then react with ozone and eventually cause a depletion of the ozone layer. Thus, since the frequency and magnitude of these events had been tied to the 11-year solar cycle, the possibility existed that the decrease observed in the late 1970s and early 1980s (as well as the reported increase in the early 1970s) was part of a natural variability of ozone associated with the solar cycle. The ozone panel was charged with the responsibility of quantifying how much of an effect, if any, solar activity had had on the observed ozone trend during this time.

In addition, there was another event of natural origin that could have had an impact on the ozone trend during this period. In 1982, the El Chichón volcano in Mexico erupted. The eruption was the largest of the decade, releasing enormous amounts of volcanic debris into the stratosphere. The eruption put more than 12 million tons of dust and gases into the stratosphere; compare this with the eruption of Mount St. Helens in 1980, which put only about a half a million tons of matter into the stratosphere (see Figure 17 in Chapter 4). When dust and accompanying gases from a volcano reach the stratosphere, particulate matter remains there for several years. Scientists know that the presence of this haze in the stratosphere causes light to be reflected differently than when volcanic matter is not present in the stratosphere. Possibly, this dust layer in the stratosphere had resulted in erroneous ozone measurements, from both space-borne and ground-based instruments. It was also possible that some of the chemicals injected into the stratosphere by the El Chichón eruption caused a depletion of the ozone in the stratosphere.

Another theory suggests that chemistry had very little to do with the presence of the ozone hole and that the very low amounts of ozone were a result of the unusual atmospheric circulation patterns that can only take place over Antarctica. These unique circulation patterns exist because the stratosphere over Antarctica during the austral spring is the coldest place on earth—by the end of the 6-month polar night the air in the lower stratosphere has become increasingly colder as it is shut off from the warming of the sun.

In contrast, at the end of the polar night at the North Pole the upper atmosphere also gets extremely cold, but not as cold as over the South Pole. The reason the air over the North Pole is not as cold is that the major mountain chains in the northern hemisphere are closer to the pole than the major mountain chain in the southern hemisphere. The presence of these large barriers deflects the relatively warm air from lower latitudes to the poles in both hemispheres. Because these deflectors are larger and closer to the pole in the northern hemisphere, the occasional deflection of the warmer air does not allow the air over the arctic to become as cold as the air over the Antarctic.

After studying all these theories and poring over the historical measurements, the panel finally released its findings in 1988. Its members concluded that the observed decrease in stratospheric ozone between 1978 and 1987 was indeed real, although they did establish that the TOMS instrument had also been undergoing a drift in its calibration since 1983. They concluded that ozone had decreased by about 3% between 1978 and 1987 in the southern hemisphere and by about 2% in the northern hemisphere. This decrease was about half of what the TOMS had measured. The panel also stated that the natural perturbations resulting from either the solar cycle or the eruption of El Chichón were not the cause of the observed decline. They concluded that the increased chlorine from CFCs in the stratosphere over the past decade had been the primary reason for the depletion of the ozone layer.[48]

While the ozone trends panel was trying to decide whether or not ozone depletion was a valid issue of concern, it was obvious that considerably more information was needed to understand why the stratosphere above the most desolate continent on earth was behaving as it was. After years of planning, scientists launched a major expedition to investigate the origin of this mysterious phenomenon that came to be known as the "ozone hole." Included in the arsenal for this 1987 scientific study were two NASA airplanes capable of sniffing and measuring trace gases. Scientists hoped the sophisticated instruments aboard these NASA airplanes would provide some of the answers to this puzzlement.

One of the NASA airplanes was a DC-8 that had been modified to accommodate more than a dozen instruments manned by nearly 50 scientists. The other was a converted U-2 spy plane that carried a few very sophisticated chemical sniffers to an altitude of almost 70,000 feet. This converted U-2, now designated ER-2 (environmental research), was capable of measuring chlorine oxide, one of the important by-products of the breakdown of the CFCs in the stratosphere.

The chlorine oxide detector was developed by Jim Anderson, an atmospheric chemist from Harvard. Fitting such a sophisticated instrument into the small confines of a converted spy plane under the hostile conditions in which the ER-2 flies can only be described as a labor of love. The detector was the result of more than a decade of intense work fueled by Anderson's dedication to a scientific idea and concept. In addition to the technical challenges of measuring, Anderson's chlorine oxide detector had to work entirely autonomously, since the ER-2 had room for only one person— the pilot. If the instrument didn't perform as it was supposed to, there was no one on board to twiddle the knobs or kick it to make it work correctly.

But the measurement of chlorine oxide was critical if scientists were to be successful in their attempts to understand why there was an ozone hole. Anderson's results from the 1987 expedition are shown in Figure 30. The measurements were obtained at an altitude of 69,000 feet as the ER-2 flew into the vortex of the ozone hole from Punta Arenas, Chile, located near the southern tip of South America. More than any other single measurement, the data obtained by Anderson's instrument confirmed the fact that the ozone hole was a manifestation of the presence of chlorine in the stratosphere: chlorine that would not have otherwise been there had it not been for the release of the CFCs many years earlier—just as Rowland and Molina had warned about in their paper in 1974.

The two traces in Figure 30 show that ozone values decreased from nearly 2.5 parts per million by volume (ppmv) to less than 0.8 ppmv as the jet traveled southward. In particular, we see the sudden drop in ozone between 70° and 71° S. Subsequent analysis of the TOMS satellite data confirmed that this was the location of the edge of the ozone hole on this particular day, marked by the latitude of the vertical line. Also shown in this figure are the measurements from Jim Anderson's chlorine oxide detector. As the plane entered the ozone hole, we see that chlorine oxide went from about 80 parts per trillion by volume (pptv) to more than 800 pptv. The smoking gun had been found.

FINGERPRINTING THE CRIMINALS

Just as every human can be identified by the unique and individual swirls and whirls of his fingerprints, every molecule in the atmosphere can be identified by its own

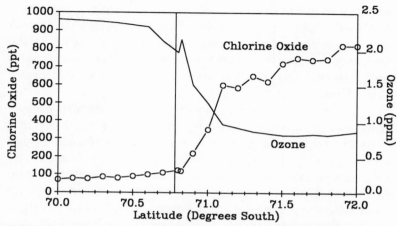

FIGURE 30. Measurements of chlorine oxide and ozone from the ER-2. Note the simultaneous decrease in ozone and increase in chlorine oxide as the ER-2 research airplane flying at 69,000 feet enters the Antarctic ozone hole, the outer boundary of which is denoted by the vertical line. The data were obtained on September 21, 1987.

unique peculiarities, which scientists call "signatures." These unique signatures are captured by the properties of the molecule's electromagnetic spectrum. Every molecule is composed of a specific number of various atoms arranged in a specific pattern. The unique combination of the atoms that make up the molecule and the way they are arranged give each particular molecule a unique set of properties. The forces that keep the molecules intact are called chemical bonds. But the bonds are not completely rigid; they can be broken apart if enough energy is applied to them. This is what happens to ozone in the stratosphere.

Ozone is made up of three oxygen atoms, held together by two fragile bonds. Ultraviolet energy from the sun, which would otherwise reach the earth's surface, can break the bonds apart. It is the same ultraviolet energy that burns our skin, causes skin cancer, destroys phytoplankton (the microscopic organisms at the bottom of the food chain) in the ocean, and that has been associated with the formation of cataracts.

It is the ozone in the upper stratosphere that prevents us from being harmed by the powerful ultraviolet energy. Nonetheless, there are other types of energy around us that are not as powerful as the harmful ultraviolet radiation. The most ubiquitous form of energy is infrared radiation, felt by us as heat. The heat that is given off by the sun intercepts the earth where it is absorbed in the ground, and then reemitted to warm the atmosphere (see Figure 23 in Chapter 7).

Heat energy (or infrared radiation) is generally not powerful enough to break a molecule apart. What can and does happen, however, is that the heat causes the bonds between the atoms to stretch. The molecule doesn't like to be stretched, so the bonds contract, and as they contract, they again release some energy. Much of this released energy is of very low frequency, often below even the infrared portion of the electromagnetic spectrum.

Microwave radiation contains less energy than is found in the infrared portion of the electromagnetic spectrum. So each molecule, if heated by infrared radiation, will radiate energy at even lower energy levels—either in the form of lower infrared radiation or even in the microwave region of the electromagnetic spectrum. But no matter where in the electromagnetic spectrum this radiation is emitted, each molecule gives back its radiation at only specific energies.

The key to the success of this technique is the ability to obtain precise measurements of the radiation by the molecules in the atmosphere so that the individual molecules can be identified and quantified. If such an instrument could be put in space looking down at the atmosphere, then scientists could determine the distribution of almost any molecule that emits infrared or microwave radiation.

MAPPING THE GUILTY PARTY

In late 1991 and early 1992, scientists launched another expedition to the northern polar regions using the DC-8 and the ER-2 NASA aircraft. The objective of the mission was to examine the chemistry at northern high latitudes, similar to the type of experiment that had been conducted for the Antarctic in 1987. Many of the participants were the same; Jim Anderson was again a major player in this campaign with his chlorine oxide detector on board the ER-2. But there was a big difference between this field campaign and any other major field experiment ever conducted previously: for the first time, there was a satellite devoted to understanding the chemistry of the stratosphere, the Upper Atmosphere Research Satellite (UARS). Looking down from its lofty perch in space UARS provides a global perspective of the measurements made by the instruments aboard the airplanes.

In one of the first flights of the field campaign—the experiment spanned five months with several flights during each month from November 1991 through March 1992—Anderson announced that he had measured concentrations of chlorine oxide twice as high as those he had measured over Antarctica in 1987. And these high levels of the ozone-eating chlorine were not seen in a region that would affect mostly penguins. These high levels of chlorine oxide were measured almost immediately after takeoff from Bangor, Maine, one of the bases of operation.

As if that news weren't bad enough, the Microwave

Limb Sounder aboard the UARS showed that even higher levels of chlorine oxide stretched from eastern Canada across the north Atlantic to the Scandinavian countries and even into parts of western Russia.[49] Thus, the ER-2 barely crossed the western boundary of a vast pool of chlorine oxide. The highest concentrations in the center of the pool exceeded 2 parts per billion, or two and a half times the 0.8 ppb (800 parts per trillion) that Anderson measured in Antarctica in 1987.

So, if there's so much chlorine in the stratosphere at northern high latitudes, why don't we see a corresponding arctic ozone hole? The answer given at the end of the 1991–1992 field mission was that temperatures don't stay cold enough for a long enough period of time for the destructive chlorine to concentrate its ozone-destroying havoc in one relatively small area. In the southern hemisphere, on the other hand, the stronger winds swirling more concentrically around the South Pole allow the temperature to get just cold enough to start a chain reaction that becomes an extremely efficient ozone destroyer. As we learn more and more about what causes the ozone hole, it appears that its creation requires the unique coincident conditions of low temperatures and complete darkness for prolonged periods followed by sunlight. As polar sunrise occurs in the early spring, the temperatures in the southern hemisphere are so cold that they remain below the threshold temperature ($-95°C$) for several weeks, resulting in the formation of certain ozone-destroying chemicals. In the northern hemisphere, on the other hand, once the sun comes up in the polar regions in February and March, the temperature is already too warm for the concentrated pools of chlorine to feast on the ozone.

Figure 31 shows the pool of ozone-destroying chlorine oxide that was measured by the Microwave Limb Sounder

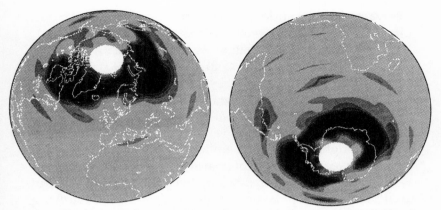

FIGURE 31. Distribution of chlorine oxide in January–February 1992 (left) and September 1992 (right). These distributions were derived from measurements obtained from the Microwave Limb Sounder aboard the Upper Atmosphere Research Satellite. (Courtesy of L. Froidevaux and J. Waters.)

shortly after UARS was launched. Even though no ozone hole per se was seen in January and February of 1992, the MLS data show that there's enough chlorine in the stratosphere to cause wide destruction of ozone at this time of the year. In fact, the Total Ozone Mapping Spectrometer found the protective ozone layer to be 10 to 15% thinner over the northern hemisphere in early 1992 than at any time during the past 13 years. The lowest levels were located over northern Europe. As if 1992 wasn't bad enough, the winter of 1993 was even worse. Not only was a substantial depletion seen at high latitudes in both hemispheres, there was now evidence, for the first time, that stratospheric ozone in the tropics had started to decrease by as much as 10% at some altitudes. The observed decrease at almost every latitude may have been catalyzed by the sudden injection of particles from Mt. Pinatubo more than a year earlier.

The complete explanation as to why such a large decrease seemed to take place in the early 1990s is still an area of intense international research. Why would the presence of volcanic particles be so important for the destruction of the ozone layer? And if Mt. Pinatubo is the reason for enhanced ozone depletion, wouldn't other volcanic eruptions throughout the past millennia also have caused a drastic decrease in the ozone layer?

The answer is, probably not, because a critical amount of chlorine must first be present before the ozone-destroying reactions that are sped up by the increased conglomeration of volcanic debris can kick into high gear. One thing that we now know is that chlorine levels in the stratosphere are now approaching 4 ppbv (parts per billion by volume) and that more than 80% of the chorine is of anthropogenic origin, the result of human activity. Thus, scientists are now fairly confident that the natural stratosphere did not contain enough chlorine to trigger the ozone destruction observed in the early 1990s.

It's somewhat ironic that when the idea for UARS was first conceived in 1976, the existence of an ozone hole had not even been considered. Neither Rowland nor Molina, who first warned us about the possible harm that inert man-made chlorine-containing chemicals could inflict on the stratosphere, had ever thought that intense ozone depletion could be concentrated in such relatively small areas. In Australia and New Zealand, advisories to avoid direct sunlight when ozone levels are low are now routinely broadcast. With the aid of satellites like UARS, the future may see ozone-hole forecasts being given days or even weeks before the ozone is destroyed in the stratosphere.

UARS is the first satellite launched in a series of earth-pointing observatories as part of NASA's Mission to Planet Earth to provide a munificence of information early in the next century. To conjecture on what new areas of research will evolve based on the measurements sent back by these sensors is not possible. But don't be surprised if we routinely hear about forecasts of depleted stratospheric ozone so that the public can take necessary precautions. Imagine how beneficial such forecasts would be. Think about how the idea of such a forecast would have been inconceivable when Fitzroy was assigned his duties in 1854, or even when Richardson began his quest to forecast the weather using human computers in the early 20th century. Indeed things have changed. And the only thing we know for certain is that they will continue to change as our level of knowledge grows.

NINE

The Revolution Has Begun

One Friday evening in March millions of Americans were surprised during the CBS Evening News when substitute anchor Bob Shieffer teased them before a commercial. Stay tuned, he told them, for a weather forecast. A weather forecast? Quite unusual. There have been weather specialists on the national network news programs. During hurricane season, Bob Sheets, director of the National Hurricane Center appears regularly to inform and entertain national audiences with the latest update about the latest hurricane bearing down on our east coast. But this is March, the hurricane season is still months away, and Bob Shieffer is about to introduce a weather forecaster on the national news. Why? What could the producers of this show be thinking?

Just before the commercial break, Shieffer introduced Dr. Louis Uccellini, chief of the Meteorological Operation Division of the National Meteorological Center in Washington, the place that feeds the rest of the country the National Weather Service forecasts. These forecasts are the basis of all the local and regional weather forecasts across the nation.

Dr. Uccellini's position is that of interpreting and modifying the raw data that go into the forecasting process.

What was he doing on national television? A glance at the national weather map didn't offer much of a clue. It showed a rather benign situation for mid-March, nothing that resembled a hurricane or blizzard, nothing that appeared as if it would compel a member of the National Meteorological Center to go on national television. Dr. Uccellini is not one of the polished, high-profile members of national government. His job is to interpret the complex information spewing forth from the gargantuan government computers singularly dedicated to providing public weather forecasts. But it was Uccellini whom audiences saw that night on television. And what he told them was simply what his computers were telling him. And what his computers told him turned out to be the weather story of the century, despite the fact that at the time of Uccellini's appearance nothing out of the ordinary had actually happened. Yet.

Dr. Uccellini appeared on the CBS National News on March 12, 1993. It was a Friday, a day when most Americans are looking forward to a two-day weekend, when interest in the weather increases. And here was a division chief in the National Meteorological Service warning them that the eastern United States needed to prepare for a storm of "historic" proportions. Even as he was speaking to the country, his agency was disseminating unprecedented forecasts of storm intensity and snowfall that, if they were correct, deserved national attention. It was going to be a disastrous weekend from Florida to Maine. On national television, a government meteorologist was predicting four feet of snow for Syracuse, New York. In the Southeast, residents could expect not only snow but possibly *tornadoes*.

Not many people watching that news program that

evening realized how unprecedented such a discussion was. It had only been about 55 years earlier that the Great New England Hurricane came ashore on Long Island in September. Without warning. Remember, that was the hurricane that managed to slip up the east coast after avoiding Miami, the hurricane that managed to travel hundreds of miles, at a rate approaching 60 mph, undetected except by one lone cruise ship.

Now, half a century later, satellites hovered far above the earth sending back pictures of the atmosphere, while sensors read humidity and temperatures; so much information, an abundance of information, fed into insatiable computers, which then spit out the future. And the future, Dr. Uccellini told the millions of viewers watching that Friday evening, looked dangerous. This was the official word from the same weather service that for decades refused to use the word *tornado* in its forecasts for fear of eliciting panic among the populace. Now a weather service spokesman was on national television predicting a storm of "historic" proportions.

Yet at the time Dr. Uccellini was making his predictions on television there was no storm pelting the United States. The weather over the eastern third of the country was rather innocuous. But Dr. Uccellini and his computers suspected as early as the previous Monday, March 8, that the following weekend would see a whopper of a storm. Remember when Lewis Richardson's prediction of a killer storm in Europe 70 years earlier proved a bust? A great deal has happened since then.

The origins of the storm that would engulf almost half the nation were relatively unremarkable little weather disturbances that appeared on weather maps as early as that Monday. A meteorologist studying the weather map on that

Monday would have no idea that a major, unprecedented storm would sweep the eastern third of the country by week's end.

If we need one example of the advances of meteorology in the past half century, we need look no further than this particular storm. Because it wasn't a meteorologist first who predicted the storm, it was a computer. Or, to be more precise, a team of computer models. What the computer models foresaw was that these minor disturbances would converge over the southeastern United States and form a massive, swirling, wind-laden storm that picked up more energy as it neared the Atlantic Ocean and tapped into the warmer air over the water before sweeping northeast. But a funny thing happened as the storm began to develop. At this point, it is important to note that Uccellini's Meteorological Operations Division uses a suite of models to offer "guidance" into the official forecasts issued by the National Meteorological Center. In general, the 3- to 5-day forecasts use models with coarser resolution than the models that are used to generate the 1- and 2-day forecasts. On March 12th, the higher-resolution computer models spewed forth 12- to 48-hour forecasts for the 13th and 14th that showed that most of the heavy snow would miss the major metropolitan areas in the Northeast. It was at this point that Uccellini's team of forecasters had to make an important decision regarding whether or not to issue a heavy snow forecast for the Northeast: do they believe what was forecast five days in advance for March 13 and 14, or do they use the guidance produced on the 12th to issue their official forecast for these days?

In the first chapter, we talked about local weather forecasting containing a certain amount of "art," especially when trying to define where it will rain and where it will snow. For accurately forecasting the March 1993 storm, the art involved understanding sophisticated computer models and why they went wrong when they did go wrong. And this, one can argue, is no longer art, but a part of the science of meteorology that allowed Uccellini and his team to stay with the heavy snow forecast despite the newly generated computer guidance that suggested otherwise.

Just how accurate were the official forecasts that came out of Uccellini's division? Syracuse, New York, did indeed set a record when it received 43 inches of snow. Mount Mitchell, in North Carolina, about 800 miles away, received 50 inches. The breadth of the storm and the intensity of it— Syracuse and Mount Mitchell receiving close to the same amount of snow yet being so distant from each other—made it truly historic. The foot of snow in northern Maine and the few inches in the Florida Panhandle were sure signs that this was, indeed, the "storm of the century." But the forecast wasn't perfect. Perhaps the biggest deficiency was the unforeseen heavy snow in eastern Kentucky and southeast Ohio, west of the Appalachian mountains when the heaviest snow was expected to stay east of the mountains. Furthermore, the record-breaking snow in northern Alabama was underpredicted. Nonetheless, the overall accuracy of the prediction was extraordinary and is an impressive milestone that would have made Lewis Richardson extremely proud.

KEEPING PLANES IN THE AIR

The fact that the "storm of the century" was monitored, predicted, and followed on the national news like any other scheduled program attests to the remarkable progress made by weather forecasters during the past several decades. The revolution has already begun, and unlike other revolutions tempered by losses in lives, this revolution in weather forecasting will be remembered for saving lives.

For example, the Federal Aviation Agency has already lent its support to the development of NEXRAD, the new generation of radars that can actually detect motions within developing thunderstorms. It didn't take much to persuade the FAA, because the agency was already concerned by a series of airplane crashes around the country that left many dead. As noted earlier, these accidents—an Eastern Airlines plane at JFK airport in 1975, a crash in New Orleans in 1982, and one in Dallas-Fort Worth in 1985—all had one thing in common: they occurred during thunderstorms and were caused by some kind of atmospheric disturbance that was not forecasted before the plane took off or landed.

The nature of the accidents was particularly frightening to pilots and navigators. There was no warning that anything outside the airplane was amiss. Most of the accidents occurred during takeoff or landing. As the plane ascended off the runway, or as it approached within feet of the surface, a sudden, ferocious downward onslaught of air would slam the plane into the ground. Even more insidious were the instantaneous rapid wind shifts that occurred with this descending air. As the rushing column of air hit the ground, the wind would diverge in all directions radiating out from the center of the shaft of the downward-moving air. And just as the pilot adjusted for the sudden shift of the wind in one direction, it shifted again in another, often totally opposite, direction. If the pilot had made proper adjustments for the first sudden change, then that same adjustment most likely turned out to be the wrong maneuver to make when the airplane encountered the second change of wind speed and direction. The result was too often disastrous.

FIGURE 32. Photograph taken at Stapleton International Airport in Denver, Colorado, during a time when a downburst from the approaching storm clouds could have resulted in an accident during landing or takeoff. (Courtesy of the National Center for Atmospheric Research.)

Despite the airport's intricate and advanced system of radars, something was hurtling the planes around like toys. And what was most frustrating was the realization that such tragedies could have been avoided had the pilots known about these hidden downbursts.

The FAA now estimates that at least 26 commercial airline accidents, responsible for more than 500 deaths, were caused by downbursts. Had there been better forecasting capabilities, these accidents could have been avoided. What are the conditions that contributed to them? Scientists now call them *downbursts*, or *microbursts* and three factors make them particularly deadly around airports.

The first is the vulnerability of the airplane as it takes off or lands. During landing and takeoff the airplane is traveling at about its slowest speed, and is only a few feet above the ground. Traveling slowly and having only a few feet of lift make the plane particularly vulnerable because it has little room to maneuver. So when a sudden shift in wind direction, or a sudden burst of intense wind comes along, there is little the plane can do to avoid it, or to adjust its flight course. The wind's effect, therefore, is to make the plane quickly lose its ability to fly.

Second, the area affected by such microbursts is so small that it is almost impossible to detect with conventional radar. None of the technological, big-ticket items such as (conventional) radars and satellites are able to detect these small bursts of energy within larger storms. Despite their being small events that would go unnoticed in most areas, they can be lethal around airports because of their effect on airplanes.

It was this danger around airports that pushed the FAA to investigate the problem. The first thing they did was to install a network of anemometers near the runways to measure wind velocity. It didn't do much good. A specially designed network of six anemometers had been installed around the Dallas–Fort Worth airport before August 2, 1985. That was the day Delta Flight 191 was coming down for a landing during a thunderstorm. Several hundred feet short of the runway, a burst of air slammed the plane into the ground, tragically killing 133 people, nearly all the passengers and crew.

There's a third factor, too. During the formation of thunderstorms, and after a well-developed thunderstorm has formed, both strong updrafts and strong downdrafts are present in addition to the precipitation and winds they produce. It is the precipitation that makes it possible to detect these storms with radar. What the radar operator sees on his console is the reflection of the radar signal bouncing off the precipitation (called an "echo") inside the thunderstorm. This precipitation can take any of a number of forms—water droplets, ice, snow—but what's important to the radar operator is that any of these forms can be detected by radar. These forms of precipitation are called *hygrometeors*. Without hygrometeors, there is nothing for the radar signal to reflect off of and so, as far as the radar operator is concerned, there is nothing out there. Since the strong downward motions of a thunderstorm often bring cold, dry air from the upper atmosphere downward, this air is often devoid of precipitation. That means that conventional radars can't always detect these areas of thunderstorms because there are little or no pieces of precipitation to generate the radar echoes indicative of severe weather.

Doppler radar, on the other hand, *can* detect motions within thunderstorms even without hygrometeors. So an obvious solution to the problem of microbursts is to equip planes with this new generation of radar. Imagine a pilot about to come in for a landing through a thunderstorm when suddenly his navigator "sees," through the radar, that there is some strong wind movement as they approach. The pilot could simply scuttle the landing pattern, circle the field, and try again. The ability to detect the invisible wind that could have slammed his plane into the ground would have saved his plane and the passengers' lives.

Back in 1986 scientists and engineers set out to see just how such a system would work. About 50 meteorologists, pilots, and aeronautical engineers gathered together for the project at NASA Langley Research Center in Hampton, Virginia, all focused on developing a method to land an airplane safely during a microburst event. The flight crew spent many hours in a training simulator specially programmed by engineers to simulate downburst conditions during which wind speed and direction change very rapidly over short periods of time. Pilots eventually developed maneuvers allowing them to avoid likely areas of a microburst averting a disastrous crash. As the time to test the new equipment and procedures approached, the crews spent several weeks flying their Boeing 737 and practicing the new maneuvers. Just this alone was quite an accomplishment.

When the pilots and engineers were ready, the group left Virginia for Florida and then Colorado to experiment with the real thing. The specially-equipped plane contained not only a Doppler radar but also a special infrared (heat) sensor that could estimate the temperature ahead of the plane. The reason for this relates to the dynamics of microbursts. A microburst is formed by a column of cool air rapidly descending through warmer ambient air. Typically, a plane penetrating through this invisible microburst would experience a warm/cool/warm temperature sequence.

A specially developed forward-looking infrared device capable of sensing temperatures up to three miles ahead was installed on the plane. The sensor was attached to an on-board computer, which then calculated the information into what scientists and engineers call the "hazard index." With a computer and sensor on board to warn the pilot of approaching hazards heretofore unseen, it was hoped that a pilot would be able to identify the deadly thermal signature and prepare himself and the plane for the microburst.

With these new instruments, crews ran experiments for over a month, both in Florida and Colorado. Mike Lewis, the NASA Langley engineer who oversaw the field tests, reported that the new remote sensors used in the experiment "acquired outstanding high-resolution measurements" of the temperature and wind fields surrounding microbursts.[50]

FIGURE 33. Photograph taken from a special camera mounted on the tail of a Boeing 737 aircraft during an experimental flight when the airplane purposely penetrated regions of suspected downburst. In two separate field experiments in 1991 and 1992 in Orlando, Florida, and Denver, Colorado, scientists and engineers flew through 75 downbursts. On board this airplane were special sensors that were able to detect the presence of a downburst several miles ahead; these remotely sensed downbursts were then verified by in situ measurements as the plane flew through them. (Courtesy of the National Aeronautics and Space Administration.)

With this kind of knowledge available to the pilot a full two minutes before actually flying through the event, he could plan for the encounter by modifying his approach. By either changing speed or altitude, pilots can avoid the deadly consequences of flying through a microburst close to the ground and at slow enough speed that the wind can toss the plane like a toy.

Doppler radar and infrared sensors are expensive, and so progress in making them a standard feature of all airplanes has been slow. But officials expect that by the end of this decade all commercial airplanes will be so equipped. That means that pilots and their passengers won't have to fear an invisible downburst while taking off or landing during thunderstorms. In the pilot's cockpit will be a small computer interpreting the information gathered by the remote Doppler radar and infrared sensors. And the pilot himself will no longer be powerless to do anything should the situation arise, because he will have been trained to respond to microburst conditions. No wonder the FAA was excited.

PUBLIC BEWARE

Back in the middle of the 19th century, when weather forecasting was in its infancy, the public was skeptical. Robert Fitzroy, a sea captain who became the first head of the Meteorological Department of Britain's Royal Society, created the first weather charts using information sent by the new-fangled telegraph from ships at sea. His forecasts were received with scorn and ridicule, not only from the public but also from his colleagues in the Royal Society. The public, being uninformed, treated Fitzroy's prognostications as sorcery. Poor Fitzroy ended up committing suicide.

Today's public, by contrast, is considerably more educated and knowledgeable about weather than the public was
in Fitzroy's time. But such enlightenment did not come
about quickly. In this country weather forecasts were first
presented on radio in 1900, as part of an experiment subsi-
dized by the Weather Bureau. Regular weather coverage,
however, did not begin until 30 years later, when weather
reports became an integral part of newscasts.[51] One of the
pioneer radio weather personalities was James Fidler, who
began broadcasting weather reports in 1934 in Muncie, Indi-
ana. Fidler's weather reports were unvarnished by humor or
gimmicks, he reported just the facts.

Weather forecasting made its television debut on New
York's experimental station (WNBT-TV) in 1941, before most
of the American public had even seen a television set. The
star of the broadcast was Wooly Lamb, a cartoon character
sponsored by Botany Wrinkle-Proof neckties in New York
City. Wooly Lamb continued on television for several years,
but most television stations featured more serious scientists
for their weathercasts.

In 1949 John Clinton Youle became the first television
weathercaster on national television, appearing nightly on
the "Camel News Caravan." Fidler moved from radio to
television as the first full-time weathercaster on NBC's "Today" show in 1952. By the mid-1950s the novelty of television
had worn off and it became an accepted part of the fabric of
American life. With competition between stations and networks increasing, broadcast executives pushed for more
entertainment on their highly rated news shows. But it was
difficult to make the hard news more entertaining, so many
broadcasters looked to the weather forecasts as the place to
inject humor.

Of course, since then we've seen the growth of so-called
"soft news," but in the early days news was seen as sacro-

sanct. The weather, then, became the repository of all kinds of goofiness. By the mid-1950s viewers had their pick of clowns, puppets, cartoon characters, animals, and scantily clad female models giving them information on the next day's weather in their area. Most of the characters were not meteorologists, and many of them didn't know the difference between a high-pressure cell and an amoeba. But they could read cue cards and they could dress up in funny costumes and makeup to make the weather funny.

This kind of treatment on national and local television dismayed professional meteorologists, for they could see the potential of this new medium to disseminate significant, even life-saving information to the general public. A rash of tornado outbreaks in 1953 and an unusual string of East Coast hurricanes in 1954, which killed hundreds of people, drove home the idea that many of these lives could have been spared had weather information been taken more seriously by television producers. After all, how much credibility was there in an image of a clown with frizzy hair and a bulbous nose telling us that there might be a tornado later that evening?

The Weather Bureau did its part to improve the way it disseminated information to radio and television stations. Meanwhile, the American Meteorological Society (AMS), a professional organization of meteorologists founded in 1919, initiated a program in 1955 to recognize those weather broadcasters who presented weather information in a dignified, credible manner. The AMS's "seal of approval" would be a stamp of credibility, assuring the viewer or listener that the person giving them weather information actually knew what he or she was talking about. The AMS's intent was to distinguish real "weather forecasters" from the television personalities who, with no background in meteorology, simply presented the weather as read off cue cards.

The first "seals of approval" were issued by the AMS in 1959. To receive the AMS seal, an applicant had to submit a film clip of his or her weather forecast to the AMS. A standing committee reviewed each submission based on informational value, educational value, audience interest, and professional attitude. In addition, the applicant's performance on television was monitored by AMS members in the viewing area over a period of time. Only after the committee and local members of the AMS agreed did the applicant receive the AMS seal of approval.

Since 1959 the AMS has granted over 700 seals of approval. The rules have evolved so that now applicants must qualify to join the Society and must take at least 12 hours of undergraduate courses in atmospheric or related science. The role of broadcast meteorology, as the AMS calls it, has evolved to a place of high esteem and importance in the fields of both meteorology and broadcasting. No longer do we see the clowns and cartoon characters and sexy babes. Instead, we see men and women trained in meteorology who have developed the ability to explain complex weather phenomena in a way average viewers can understand.

The importance of broadcast meteorology among professional meteorologists is evidenced by the recent election of Bob Ryan as the AMS president. Ryan was once the weatherman for NBC's "Today" show, later becoming the broadcast meteorologist for station WRC-TV in Washington, D.C. Ryan is the first AMS president to be elected from the ranks of broadcast meteorology. Prior to his election, every AMS president came from the ranks of researchers, professors, or government administrators.

By the 1970s the image of the broadcast meteorologist had changed. Stations found it to be a competitive edge to have an AMS-certified meteorologist giving the weather report in their market. The public responded by changing their attitude and taking the weather reports more seriously. Weather forecasters on television became popular on their own merits, and stations spent time and money promoting them to the public. This change in attitude was helped along by changes in technology that enabled broadcast weather forecasts to be understood in entirely different ways than was previously possible.

By the 1970s, NOAA's geostationary satellites brought quality images of the earth viewed from space to television stations everywhere. More and more stations invested in their own radar and computer equipment. In the old days, a typical weather forecast on television featured a blank map of the United States, on which the weather forecaster drew the fronts and isobars and storm systems with a grease pen. Significant weather was shown as a paste-up umbrella or a snowflake. He or she wrote in some sample temperatures and that comprised the national weather scene.

Compare that with today's animated, realistic views of the weather changing before your eyes, with satellite images beamed from space and looped time-series full color images giving you, the viewer, undreamed-of visual knowledge of how weather is made. What viewers see today on their television screens is the result of a major breakthrough in 1981 by Colorgraphics Systems in Madison, Wisconsin— a company that commercialized many of the automated analysis systems that had been developed in the Space Science and Engineering Center at the University of Wisconsin in Madison. Many of these innovations at the university were initiated by Professor Verner Suoml, one of the pioneers in the field of satellite meterology. What Colorgraphics did was to package a satellite-depiction and graphics system that individual television stations could use to enhance their daily weather broadcasts. The new development caught on like wildfire, and by the mid-1980s almost every television station had the capability of producing a high-tech, computer-generated, accurate weather forecast. The weather broadcast in a small market like Fargo, North Dakota, could have as professional and modern a look as a station in New York or Los Angeles.

Today's computer-driven graphics packages allow the local forecaster to superimpose fronts, jet streams, and temperature contours in such a way that the explanation of weather can be as entertaining as some of the entertainment programs, but without sacrificing good science, good education, and good taste.

REVOLUTION VERSUS EVOLUTION

Good science boils down to making as many high-quality observations as possible, and then analyzing and interpreting them. During the past century we've witnessed our ability to make high-quality observations increase at a rate that sometimes outpaced our ability to analyze and interpret. Now, with new computer technology, we are catching up to a point where the data fed from our space-borne observation systems can be digested by computer systems not even dreamed of a mere twenty years ago.

By the end of this century we'll have satellites up in space with even more capability. The new GOES satellite is only the first of a new generation of satellites that promise to supply us with trillions of bytes of information we would not have been able to process until now. On the drawing board are space-borne platforms that will provide specialized data bases that will help us to understand atmospheric processes to a degree only hypothesized a decade ago. We'll know exactly where a cloud will form before it can even be seen. Vast regions of data voids will cease to exist. Ironically, because of the relative constancy of the ocean surface, satellites will actually be able to obtain more accurate information over water than over land surfaces.

A new generation of meteorologists will be entering the profession by the next century. They will not have known the practice of meteorology before the computer and satellites. These scientists will be the ones responsible for the interpretation and analysis of the GOES-NEXT series of satellites, the Earth Observing System, and the first lasers in space. They will lead us to a new understanding of the weather, and to a day when an accurate forecast for next week, next month, even next year will be taken for granted by the public.

Imagine it is March, and you're planning a birthday party for June. Should you plan a day at the beach, or some kind of indoor activity? In the future, you'll be able to count on an accurate weather forecast three months in advance. Impossible? It was once thought impossible to forecast hurricanes. Time will tell.

Endnotes

1. C. S. Ramage, "Forecasting in Meteorology," *Bulletin American of the Meterological Society* 74 (October 1993), 1863–1871.
2. Ibid.
3. Ibid.
4. Robert J. Curran *et al.*, *LASA, Lidar Atmospheric Sounder and Altimeter*, Instrument Panel Report, Vol. IId (Washington, D.C.: NASA, 1987).
5. Edward N. Lorenz, *The Nature and Theory of the General Circulation of the Atmosphere*, WMO—No. 218. TP. 115, World Meteorological Organization, Geneva: 161 pp., 1967.
6. Vilhelm Bjerknes, "Das Problem der Wettervorhersag betrachtet vom Standpunkte der Mechanik und der Physik," *Meteor. Zeit* 21 (1904), 1–7.
7. Frederick G. Shuman, "History of Numerical Weather Prediction at the National Meteorological Center," *Weather and Forecasting* 4 (September 1989), 286–296.
8. Lewis F. Richardson, *Weather Prediction by Numerical Process* (Cambridge, England: Cambridge University Press, 1922).
9. Ibid.
10. Richardson served in World War I as an ambulance driver since he declared himself a pacifist—having been raised as a Quaker.

11. Shuman, 1989, pp. 286–296.
12. Ibid.
13. An interesting discussion behind the development of weather satellites can be found in "A History of Civilian Weather Satellites," in *Weather Satellites: Systems, Data and Environmental Applications* (P. K. Rao et al., eds.), (Boston: American Meteorological Society, 1990), pp. 7–19.
14. M. P. McCormick *et al.*, "Scientific Investigations Planned for the Lidar In-Space Technology Experiment (LITE)," *Bulletin of the American Meteorological Society* 74 (February 1993), 205–214.
15. Edward V. Browell *et al.*, "NASA Multipurpose Airborne DIAL System and Measurements of Ozone and Aerosol Particles, *Applied Optics* 22 (1983), 522–534.
16. An excellent summary of the history of hurricanes can be found in Joe McCarthy, *Hurricane!* (New York: American Heritage Press, 1969).
17. Robert C. Sheets, "The National Hurricane Center–Past, Present, and Future," *Weather and Forecasting* 5 (June 1990), 185–232.
18. Bob Case, "Hurricanes: Strong Storms Out of Africa," *Weatherwise* 43 (February 1990), 23–29.
19. Rick Gore, "Andrew Aftermath," *National Geographic* 183 (4) April 1993, 2–37.
20. Ibid.
21. Gray's scientific hypothesis, based on an examination of historical weather data, was published in William M. Gray, "Atlantic Seasonal Hurricane Frequency. Part I: El Niño and 30 mb Quasi-Biennial Oscillation Influences," *Monthly Weather Review* 112 (September 1984), 1649–1668. A subsequent analysis describing, among other things, the success of his forecasting technique, is discussed in William M. Gray *et al.*, "Predicting Atlantic Seasonal Hurricane Activity 6–11 Months in Advance," *Weather and Forecasting* 7 (September 1992), 440–455.
22. A fascinating memoir has been written by Professor Fujita, *The Mystery of Severe Storms, During the 50 Years, 1942–1992,*

Wind Research Laboratory Research Paper 239, The University of Chicago, Chicago, 1992, 298 pp.

23. Samuel Milner, "NEXRAD—The Coming Revolution in Radar Storm Detection and Warning," *Weatherwise* 39 (April 1986), 72–85.

24. Climate definitions can traditionally be found in a host of general meteorology texts. See, for example, Chapter 17 in S. Petterssen, *Introduction to Meteorology* (3rd ed.), (New York: McGraw-Hill), 1969 pp. 276–288.

25. In *The Genesis Strategy* (New York: Plenum, 1976), 419 pp., Stephen H. Schneider often refers to Bryson's work to warn of the potential of global cooling.

26. J. T. Houghton, G. J. Jenkins, and J. J. Ephraums, eds, *Climate Change, the IPCC Scientific Assessment* (Cambridge, England: Cambridge University Press, 1990).

27. Ibid.

28. *New York Times*, June 24, 1988.

29. G. T. Walker, "Correlation in Seasonal Variations of the Weather. Part Viii: A Preliminary Study of World Weather," *Memoirs of the Indian Met. Dept.* 24 (1923), 75–131, and G. T. Walker and E. W. Bliss, "World Weather V," *Mem. Royal Met. Soc.* 4 (1932), 53–84.

30. For example, see discussion by H. F. Diaz and G. N. Kilndos, "Atmospheric Teleconnections Associated with the Extreme Phases of the Southern Oscillation," in *El Niño: Historical and Paleoclimatic Aspects of the Southern Oscillation* (Henry F. Diaz and Vera Markgraf, eds.), (Cambridge, England: Cambridge University Press, 1992), pp. 7–28.

31. S. George Philander, *El Niño, La Niña, and the Southern Oscillation* (New York: Academic Press, 1990).

32. William H. Quinn, "A Study of Southern Oscillation-Related Climatic Activity for A.D. 622–1900 Incorporating Nile River Flood Data," in *El Niño: Historical and Paleoclimatic Aspects of the Southern Oscillation*, pp. 119–149.

33. Ibid.

34. Ibid.
35. Ibid.
36. Gerald A. Meehl and Grant W. Branstator, "Coupled Climate Model Simulation of El Niño/Southern Oscillation: Implications for Paleoclimate," in *El Niño: Historical and Paleoclimatic Aspects of the Southern Oscillation*, pp. 69–92.
37. J. T. Houghton, G. J. Jenkins, and J. J. Ephraums, eds., *Climate Change, the IPCC Scientific Assessment* (Cambridge, England: Cambridge University Press, 1990).
38. R. J. Charlson *et al.*, "Perturbation of the Northern Hemisphere Radiative Balance by Backscattering from Anthropogenic Sulfate Aerosols," *Tellus* 43AB (1991), 152–163.
39. J. Fishman, "Probing Planetary Pollution from Space," *Environ. Sci. Tech.* 25 (4) (1991), 612–621.
40. A. Volz and D. Kley, "Evaluation of the Montsouris Series of Ozone Measurements Made in the Nineteenth Century," *Nature* 332 (1988), 240–243.
41. Samuel J. Oltmans and Hiram Levy II, "Surface Ozone Measurements from a Global Network," *Atmospheric Environment* 28 (1994), 9–24.
42. J. Fishman, "The Global Consequences of Increasing Tropospheric Ozone Concentrations," *Chemosphere* 22 (1991), 685–695.
43. Jennifer A. Logan, "Tropospheric Ozone: Seasonal Behavior, Trends, and Anthropogenic Influence," *J. Geophys Res.* 90 (1985), 463–482.
44. P. Minnis *et al.*, "Radiative Climate Forcing by the Mount Pinatubo Eruption," *Science* 259 (5 March 1993), 1411–1415.
45. J. C. Farman, B. G. Gardiner, and J. D. Shanklin, "Large Losses of Total Ozone over Antarctica Reveal Seasonal ClO_x/NO_x Interaction," *Nature* 315 (1985), 207–210.
46. *Scientific Assessment of Ozone Depletion: 1991*, World Meteorological Organization Global Ozone Research and Monitoring Project—Report No. 25 (Geneva: World Meteorological Organization, 1991).
47. L. B. Callis and M. Natarajan, "The Antarctic Ozone Mini-

mum: Relationship to Odd Nitrogen, Odd Chlorine, the Final Warming, and the 11-year Solar Cycle, *J. Geophys. Res.*, 91, pp. 10,771–10,796, 1985.

48. *Scientific Assessment of Ozone Depletion: 1991*, World Meteorological Organization Global Ozone Research and Monitoring Project—Report No. 25.

49. J. W. Waters *et al.*, "Stratospheric ClO and Ozone from the Microwave Limb Sounder on the Upper Atmosphere Research Satellite," *Nature* 362 (15 April 1993), 597–602.

50. Michael S. Lewis *et al.*, "Design and Conduct of a Windshear Detection Flight Experiment," in *Society of Flight Test Engineers 24th Annual Symposium*, Society of Flight Test Engineers, 1993, Lancaster, CA: pp. 6-21–6-38.

51. Ben Gelber, "Television's Changing Climate," *Storm* 1 (August 1993), 32–35.

Index